GALAXIES
AND THE UNIVERSE

An observing guide from DEEP SKY magazine

Edited by David J. Eicher

Associate Editor, ASTRONOMY

With contributions by Robert Bunge, Jeffrey Corder, Alan Goldstein, Steve Gottlieb, David Higgins, Alister Ling, Tom Polakis, Max Paul Radloff, and Brian Skiff

Kalmbach Books
Waukesha, Wisconsin
1992

Books by David J. Eicher

The Universe from Your Backyard; a guide to deep-sky objects from ASTRONOMY magazine (Cambridge University Press and AstroMedia, New York, 1988)

Deep Sky Observing with Small Telescopes; a guide and reference (Editor and coauthor; Enslow Publishers, Hillside, New Jersey, 1989)

Beyond the Solar System; 100 best deep-sky objects for amateur astronomers (AstroMedia, Waukesha, Wisconsin, 1992)

Stars and Galaxies; ASTRONOMY's guide to exploring the cosmos (Editor and coauthor; AstroMedia, Waukesha, Wisconsin, 1992)

The New Cosmos (Coauthor; Kalmbach Books, Waukesha, Wisconsin, 1992)

Galaxies and the Universe; an observing guide from **Deep Sky** ***magazine*** (Editor; Kalmbach Books, Waukesha, Wisconsin, 1992)

FOR JOHN HAROLD EICHER,
without whom there would have been no *Deep Sky* magazine.

Art director: Lawrence Luser
Designer: Patti Keipe

The material in this book first appeared as articles in Deep Sky magazine. They are reprinted here in their entirety.

© 1992 by Kalmbach Publishing Co. All rights reserved. This book may not be reproduced in part or in whole without written permission from the publisher, except in the case of brief quotations used in reviews. Published by Kalmbach Publishing Co., 21027 Crossroads Circle, P.O. Box 1612, Waukesha, WI 53187. Printed in U.S.A.

Library of Congress Cataloging-in-Publication Data

    Galaxies and the universe: an observing guide from Deep Sky magazine / edited by David J. Eicher: with contributions by Robert Bunge . . . [et al.].
      p.   cm.
    Includes bibliographical references and index.
    ISBN 0-913135-14-3
    1. Galaxies. I. Eicher, David J., 1961-   . II. Bunge, Robert. III. Deep sky.
QB857.G379   1992   92-20539
523.1'2–dc20

Contents

Preface by David J. Eicher 5

A World of Galaxies
Observing the Local Group of Galaxies by Tom Polakis 6
Observing the Morphology of Galaxies by Alan Goldstein 14
Observing Interacting Galaxies by Alan Goldstein 22
Observing the M81 Galaxy Group by Tom Polakis 30

A Color Portfolio of Galaxies 38

The Biggest and Brightest
All about M31 by Brian Skiff 54
Exploring the Region of M51 by Robert Bunge 62
The Galaxies of Canes Venatici by Max Paul Radloff 68
NGC 7331 and its Ambiguous Galaxies by Jeffrey Corder 76

Galaxies to Challenge You
Dwarf Galaxies for "Dwarf" Telescopes by Alister Ling 82
A Trio of Springtime Galaxy Groups by Jeffrey Corder 84
The M31 Globular Cluster System by David Higgins 90
The Galaxies of Orion by Steve Gottlieb 96
The Coma Berenices and Abell 1367 Galaxy Groups by Brian Skiff ... 98
A Night of Galaxies near M13 by Robert Bunge 106

Bibliography 110
Index 111

Photo by Paul Roques

Preface

In June 1977 I began publishing a mimeographed newsletter for amateur observers of nebulae, clusters, and galaxies. Unceremoniously titled *Deep Sky Monthly*, the little publication began with a circulation of 25 and throughout its five-year existence grew to about 1,000. Along the way contributors sent articles and photographs, and the newsletter became a magazine, sporting glossy paper and a saddle-stitched binding by early 1982. Later that year the publishers of ASTRONOMY bought the journal, hired me as its editor (and as an editor of ASTRONOMY), and reissued the publication as the quarterly *Deep Sky*. The little magazine grew from 1,000 to 13,500 in circulation and survived nearly 10 years as the most successful specialized magazine in amateur astronomy. Times changed and priorities did too, however, and by late 1991 Kalmbach Publishing Co., the owner of *Deep Sky*, decided to stop its publication.

In its nearly 15-year lifetime, the little magazine racked up some pretty impressive statistics. Altogether, we published 1,037 articles, 2,059 photographs, and 900 sketches over the lifetime of the magazine. *Galaxies and the Universe* represents some of the best material assembled during the existence of *Deep Sky* magazine. Its fourteen articles will take you out under the stars and explore with you the world of galaxies as they appear from your backyard. The galaxies included range from the biggest and brightest — like the Andromeda Galaxy, M33, and M51 — to the most challenging, like Maffei 1 and the Coma Berenices galaxy cluster. Whether you have a 4-inch reflector or a 25-inch Dobsonian, this book will offer you plentiful observing opportunities during all four seasons of the year.

The writers of this book represent some of the brightest stars in the world of amateur astronomy. Robert Bunge is an editor who lives in Laurel, Maryland, whose extensive observational experience dates to his days using the 31-inch reflector at Warren Rupp Observatory near Mansfield, Ohio. Jeffrey Corder is a botanist in Florida whose observational skills center on galaxy clusters and astronomical sketching. Alan Goldstein is a cofounder and national coordinator of the National Deep Sky Observers Society, a Louisville-based group of galaxy watchers. Steve Gottlieb is a California math instructor with years of experience in observing galaxies and computer-based astronomy. A resident of El Paso, Texas, David Higgins has owned several large telescopes and is one of the most active supporters of the Texas Star Party. Alister Ling is a meteorologist in Edmonton, Alberta, with combined passions of daytime and nighttime skywatching. A Phoenix-based engineer, Tom Polakis has written articles with technically precise analyses of recent data. Max Paul Radloff is a professional musician in Saint Paul, Minnesota, whose observational interests center on galaxies. Brian Skiff is an astronomer at Lowell Observatory in Flagstaff, Arizona.

Although *Deep Sky* magazine no longer exists, extensive coverage of deep-sky observing can be found in ASTRONOMY magazine, the largest and most lavishly illustrated periodical of its type. For more information on ASTRONOMY, check your local newsstand or write to Kalmbach Publishing Co., 21027 Crossroads Circle, Waukesha, Wisconsin 53187. Happy observing!

Dave Eicher
Waukesha, Wisconsin
April 1992

Observing the Local Group

An exploration through our own galaxy cluster
by Tom Polakis

One of the most fascinating properties of the universe is that gravity applies at all scales. This can be seen from the Moon's lock into an orbit about Earth all the way up to the arrangement of galaxies in superclusters. Amateurs often observe galaxy clusters, from the nearby Virgo cluster out to distant groupings such as the Corona Borealis cluster. Our own Milky Way galaxy is a member of the galaxy cluster called the Local Group.

The center of the Local Group lies approximately between the Milky Way and the Andromeda Galaxy, M31. Extending around this center to a radius of about 1.5 megaparsecs is a cluster of over thirty galaxies, most of them within reach of amateur telescopes. The great bulk of the mass of the Local Group is contained within the Milky Way and M31. These two galaxies, in fact, are massive enough to have their own satellite galaxies. The Milky Way has eleven known companions and M31 sports seven.

What makes a galaxy a member of the Local Group? The best discussion of this subject, although somewhat technical, appears in *The Astrophysical Journal.* Authors Amos Yahil, Gustav Tammann, and Allan Sandage (1977) build on Edwin Hubble's observation that Local Group members do not partake in the general cosmological expansion seen everywhere else in intergalactic space. The authors use existing radial velocity data to define membership on kinematic grounds, rather than just arbitrarily stating "it's nearby, so it's a member." For membership in the Local Group, a galaxy must be moving with our cluster independent of the "Hubble flow" (expansion). By plotting the spatial distribution of the galaxies, the diameter of the Local Group was shown to be no less than 3 Mpc.

The list of Local Group members here is adapted from Paul Hodge's book *Galaxies,* with a few additions of recent discoveries made since the book's publication in 1986. This list is the most consistent with the definition of Local Group membership. Note the absence of some pretty well resolved nearby galaxies like Maffei I, NGC 247, and Sextans A, which have been denied membership in recent years based on their measured velocities.

For the visual observer, the Local Group galaxies are interesting because we can see features commonly observed in our own Galaxy, such as nebulae and star clusters. Although all but three of the members are elliptical or irregular, there is a wide variety of galactic structure to be observed. Because of their extremely low surface brightness, spotting many of these objects provides the utmost challenge to amateurs. Paradoxically, these are also the closest members, the dwarf ellipticals.

Most Local Group galaxies can be found in the evening autumn sky. Since we are dealing with some difficult objects, the most careful practices of deep-sky observing are required. Full dark-adaptation and a dark sky are required to see these galaxies in any detail. I found that slowly sweeping the telescope across the objects helped in first detecting them. A good wide-field eyepiece giving an exit pupil of at least 5 mm will help, too. One important observing accessory is a dark cloth to block out any local lights or skyglow, available at fabric stores for a few bucks. Although this antisocial behavior seems a little strange at star parties, several minutes under the monk's hood assists dramatically on faint objects.

The Local Group galaxies in this article are arranged by type. We'll start our tour with the companions to the Andromeda Galaxy, then take a look at the irregulars and remaining ellipticals of our galaxy cluster. The "big four" (M31, M33, and the Magellanic Clouds) are too detailed to be covered here. M31 has been covered in previous issues of this magazine by Brian Skiff (1984) and by David Higgins (1990). An article by Skiff (1983) covers M33 in detail.

The Andromeda Companions

Let's start out with M31's seven companion galaxies. In the same low power field with the Andromeda Galaxy — and familiar to every amateur — are **NGC 205** and **M32** (NGC 221). Imagine the view of M31 from one of those galaxies. The sky would be filled with this spectacular spiral, blocking any of the sky behind it with its stellar associations and dust clouds. The Andromeda Galaxy has made its presence known gravitationally as well because both NGC 205 and M32 show evidence that their outer regions have been distorted by tidal forces. In my 13-inch f/4.5 Newtonian from a dark Arizona site, NGC 205 shows a 9' by 5' halo, oriented in position angle (p.a.) 150°, which is steadily brighter to the central arcminute, where a bright core blazes. There is a 12th-magnitude star halfway out to the south end of the galaxy's face. M32 is always obvious next to M31, even from urban observing sites. It is quite small, only 3' by 2', elongated north-south, with a very bright center and a rapid fall-off to the sky background. It takes high power well, showing a 7th-magnitude star 12' away in p.a. 30° and several other 9th-magnitude field stars within one-half degree.

Detached from M31 and its two immediate companions are **NGC 147** and **NGC 185**. These serve as a preview of the

The elusive galaxy NGC 6822 in Sagittarius is one of the splendors of our Local Group. Four major HII regions are visible in this hydrogen-alpha image. Are these visible in backyard telescopes? Photo courtesy Paul Hodge.

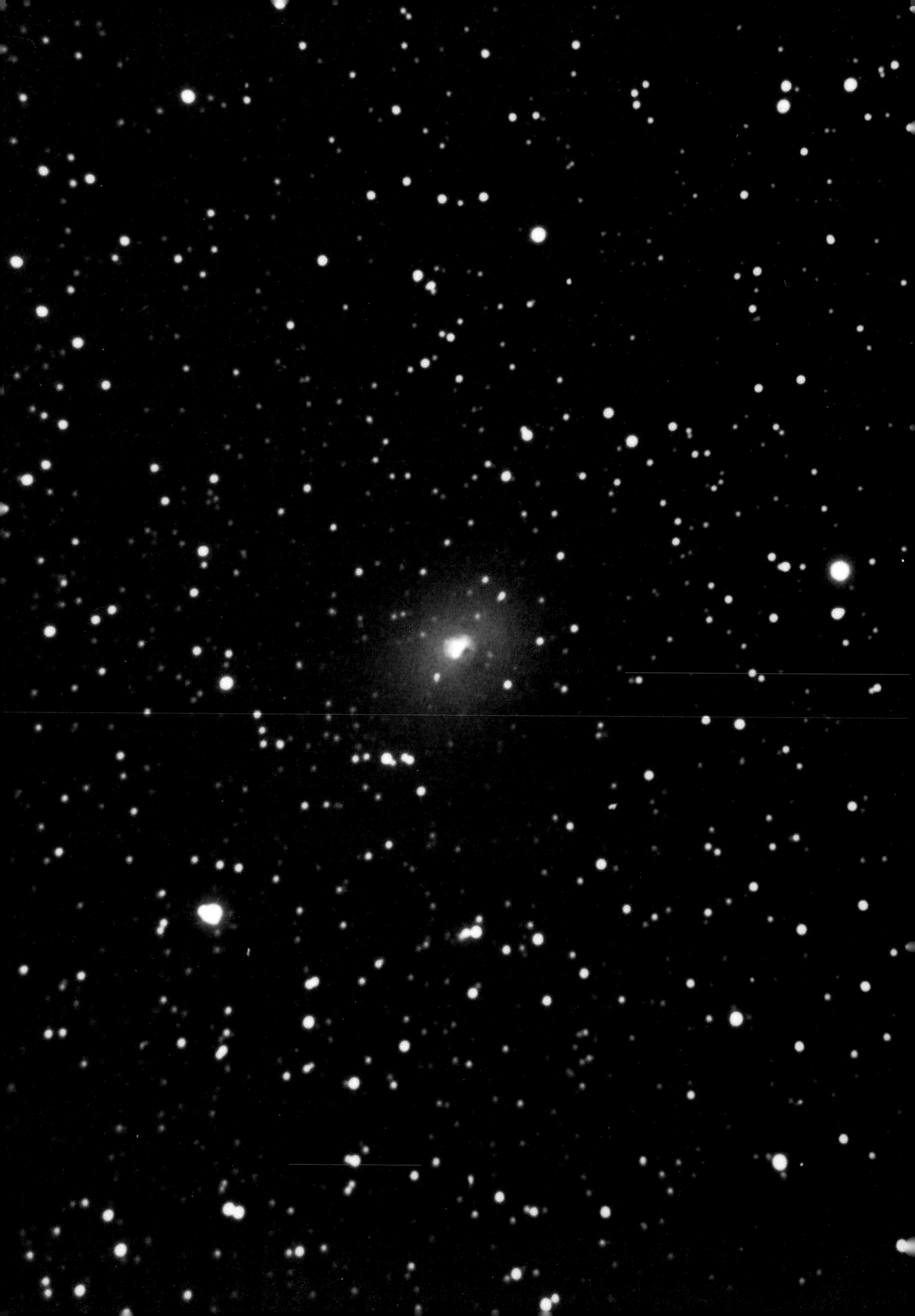

typical appearance of Local Group galaxies: dwarfs of low surface brightness. The two are close enough in space that they are likely gravitationally bound. In the 13-inch, NGC 147 appeared to be 8' by 4' in size although its published size is 13' by 8'. This is quite common as the professional definition of "size" is often for a lower level of brightness than can typically be seen in amateur scopes. The galaxy appears nearly uniform at first glance but begins to show some central brightening after a while. The field at this low galactic latitude is peppered with many stars, including a 13th-magnitude star embedded in the glow just west of the nucleus. One degree to the east, NGC 185 can be squeezed into the same low power field with NGC 147.

It is more easily seen than the latter, showing a 5' by 3' halo around a bright center, similar to a spiral galaxy. By cranking the power up to 215x, however, there was no sign of the stellar nucleus we'd see if it were actually a spiral. This magnification began to reveal a grainy texture to NGC 185, unlike its companion to the west. There are five known globular clusters in this galaxy (Hodge, 1974). At high power (310x), I was able to see globulars #1 and #2, with #2 being more visible, even appearing nonstellar. Globular #1 was stellar and visible only about 10 percent of the time.

Visibility of the Andromeda Galaxy's companions drops off quickly after these four bright ellipticals. **Andromeda I**, **Andromeda II**, and **Andromeda III** are exceedingly faint dwarf elliptical galaxies discovered by Sidney van den Bergh (1972) in a search of 48-inch Schmidt plates. These plates, which were of higher quality than those from the *Palomar Observatory Sky Survey* (*POSS*), were taken in 1970 specifically to search for companions to M31. Andromeda I and II were later verified by their appearance on the *POSS*, but Andromeda III remains invisible on these plates. Also found on the 1970 plates was **Andromeda IV**, but it was soon dismissed as a non-member of the Local Group. Visually, only Andromeda I was readily seen in my 13-inch as a 4' diameter, round glow of uniform surface brightness, just above the brightness level of the night sky. Eight 12th- and 13th-magnitude stars appeared over its face in a small group. During several persistent attempts, I was not convinced I saw either Andromeda II or III. Andromeda II is situated behind a faint Milky Way star cloud which gives some false hope. It would be interesting to know if larger apertures (24-inch?) can dig up these galaxies.

The Irregulars

Perhaps the most interesting members of the Local Group are its dwarf irregulars. These often show the same features found in spiral galaxies such as H II regions, star clouds, and dark nebulae. Again, this class of galaxies runs the gamut of brightness. In Sagittarius alone they range from the much-observed NGC 6822 to SagDIG, a nearly invisible dwarf.

E.E. Barnard discovered the first entry on our list with his 5-inch refractor in 1884. **Barnard's Galaxy**, now known as NGC 6822, was important in Edwin Hubble's early work with determining the distance scale to these types of "nebulae." Hubble's 1925 paper was the first to demonstrate the use of the period-luminosity relationship of Cepheid variables to regions outside our own Galaxy, making this the first object to be shown conclusively to be extragalactic. And Barnard's Galaxy has been much studied since this time. Paul Hodge (1977) found 26 additional star clusters in addition to the original five found by Hubble. He also catalogued eleven new dark nebulae and added eleven H II regions to Hubble's original list of five. Several of these, we will see, are visible in modest-aperture scopes.

The overall appearance of NGC 6822 is of a pretty even 9' by 4' glow elongated north-south. A faint "bar," much discussed in the literature, is located almost centrally and is elongated 4' by 1' along the galaxy's major axis. With his 13-inch f/5.6 telescope at 60x, Steve Coe saw this galaxy elongated in a 3:2 ratio with some central brightening and a mottled appearance. I counted 12 stars across the face, the brightest glowing at 12th magnitude on the northern end. While I was unable to discern any of the dark clouds or star clusters, the 13-inch easily showed two H II regions and a third, more difficult nebulous patch. The drawing shows the three H II regions, all at the galaxy's northern end. The easternmost, Hubble No. 14, "blinked" well with an O III filter and showed a round, uniform disk 15" across at 310x. Hubble No. 9, in the center, is brighter than No. 14, and shows some brightening at its center. Finally, Hubble No. 4 is the most difficult, a round, 30"-diameter diffuse patch around a tiny clump of stars, not responding at all to the O III filter.

IC 1613 in Cetus sits at a similar distance as NGC 6822. This irregular galaxy shows very little in the way of dust and few bright open clusters (although 25 quite small ones have been found). CCD images taken with the Kitt Peak 2.1-m telescope through an H-alpha filter (Hodge, Lee and Gurwell, 1990) have brought the number of H II regions to 77. Visually, IC 1613 is a fairly difficult object, showing the lack of contrast with the night sky so common in the Local Group galaxies. At a low power of 75x, it appears perfectly round, a uniform brightening of a poor star field. To make viewing more difficult, an 8th-magnitude star 10' to the northwest almost spoils the view. Most of the more prominent H II regions are in the northwest quadrant of the galaxy, but the 13-inch failed to show them.

Moving back toward the galactic plane, we find **IC 10** in Cassiopeia. This galaxy shows a high B-V value, being strongly reddened due to obscuration by our own Milky Way. IC 10 is positioned where a galaxy has no business showing up: only 3° from the galactic plane. Therefore, it is pretty difficult to observe, despite its close proximity. The 13-inch showed a faint, 4' by 3' mist against a bright background, elongated northwest-to-southeast. The star field here is incredibly rich and reminded me of the field around the beautiful edge-on galaxy in Andromeda, NGC 891. A 12th-magnitude star shines in front of its center with two other 13th-magnitude stars to the north, still in the small glow of the galaxy. Although the detail is not spectacular, this one exhibits that magnificent "three dimensional" effect of a galaxy forming a backdrop to a Milky Way starfield.

The autumn night sky contains six more Local Group irregulars. Surprisingly bright is **WLM**, named after astronomers Max Wolf, Knut Lundmark, and P.J. Melotte, who discovered it early in this century. P.J. Melotte (1926) commented on the striking similarity between this galaxy and NGC 6822. Its stars, of 18th magnitude, are easily resolved on plates, and it even contains one globular cluster. Don't be daunted by the low surface brightness given in the table as WLM was pretty easy in the 13-inch. It was elongated 9' by 4' north-south, and appeared uniform across most of this area. The outer 1' around its periphery showed a rapid fall-off to the sky brightness. Three stars are easily seen superposed on the galaxy's face.

At a distance of nearly 1,750 kpc, the **Pegasus Dwarf** is truly at the edge of the Local Group, near its "zero velocity surface." It appears 8' by 3' in extent, elongated in p.a. 105°, and pretty faint overall but with some central brightening. There are eight bright stars across the face of the galaxy in what is otherwise a poor starfield. About 20' south is a chain of six stars in an arc ending at a 9th-magnitude star.

Local Group System 3 is a tough object, discovered in

NGC 185 in Cassiopeia is one of the two challenging satellites to M31. Are the dust patches in this galaxy visible in backyard scopes? Photo by Harvey Freed (10-inch SCT at f/6.3, hypersensitized Tech Pan film, 45-minute exposure).

Galaxies in the Local Group

Name	Other	Type	R.A. (2000.0) Dec.	Mag (V)	SB (V)	Size	Dist.	Con.
WLM		IB(s) IV-V	0h02.0m, -15°28'	11.0	14.9	10'.2 by 4'.2	960 kpc	Cet
IC 10		KBm?	0h20.4m, +59°18'	10.3	13.5	5'.1 by 4'.3	870	Cas
NGC 147		dE5 pec	0h33.2m, +48°31'	9.5	14.5	13' by 8'	610	Cas
And III		—	0h35.4m, +36°31'	—	—	4'.5 by 4'.0	—	And
NGC 185		dE3 pec	0h39.0m, +84°20'	9.2	14.3	11' by 10'	570	Cas
NGC 205	"M110"	E5 pec	0h41.3m, +41°41'	8.0	13.6	17' by 10'	760	And
NGC 221	M32	cE2	0h42.7m, +40°52	8.2	12.3	7'.6 by 5'.8	760	And
NGC 224	M31	SA(s)b 1-11	0h42.7m, +41°16'	3.5	13.4	178' by 63'	760	And
And I		dE3 pec?	0h45.7m, +38°00'	13.2	16.2	3'.5: by 2'.5:	—	And
SMC		SB(s)m pec	0h51.7m, -73°14'	2.3	13.8	4°.7 by 2°.6	60	Tuc
Sculptor	E351-G30	dE3 pec	1h00.0m, -33°42'	9.2	16.7:	28' by 23'	110	Scl
LGS a3	Pisces	—	1h03.8m, +21°53'	—	—	—	—	Psc
IC 1613	DDO 8	IAB(s)m V	1h05.1m, +2°08'	9.4	14.6	12' by 11'	730	Cet
And II			1h16.4m, +33°27'	<15		3'.5 by 3'.5	730	Psc
NGC 598	M 33	SA(s)cd II-III	1h33.9m, +30°39'	5.7	14.1	62' by 39'	1100	Tri
Formax	E356-G04	dE2	2h93.9m, -34°32'	9.0(B)	15.9(B)	12' by 10'	130	For
UGC-A86		—	3h59.9m, +67°08'	—	—	—	630	Cam
LMC		SB(s)m	5h19.7m, -68°57'	0.1	13.9	10°.8 by 9°.2	54	Dor
Carina	E206-G220	dE3	6h14.6m, -50°58'	20.4:	—	35' by 26'	85	Car
Leo A	Leo III	IBm V	9h59.4m, +30°45'	12.6	15.5	4'.9 by 3'.2	1600	Leo
Leo I	Regulus Sys	dE3	10h05.5m, +12°19'	10.1	15.0	10'.7 by 8'.3	230	Leo
Sextans		dE3	10h13.2m, -1°37'	11	20	90' by 65'	85	Sex
Leo II	Leo B	dE0 pec	11h13.5m, +22°10'	11.5	17.1	14' by 13'	230	Leo
GR 8	DDO 155	Im V	12h58.7m, +14°13'	14.4:	14.6:	1'.2 by 1'.1	1000	Vir
Ursa Minor	DDO 199	dE4	15h08.8m, +67°12'	<14	—	32' by 21'	67	Umi
Draco	DDO 208	dE0 pec	17h20.1m, 57°55'	<14	—	40' by 25'	75	Dra
Sag DIG	E594-G04	IB(s)m V	19h30.1m, -17°42'	—	—	3'.3 by 1'.8	—	Sgr
NGC 6822	Barnd's Gal.	IB(s)m IV-V	19h44.9m, -14°49'	8.6:	13.5:	10'.2 by 9'.5	620	Sgr
Aquarius	DDO 210	Im V	20h46.8m, -12°51'	14.7	15.4	2'.1 by 1'.2	—	Aqr
IC 5152		IAB(s)m IV	22h06.1m, -51°17'	10.5	13.1	4'.9 by 2'.8	1500	Ind
Tucana		dE5	22h41.7m, -64°25'	15.0:	17.5:	5' by 2'.5	290	Tuc
Pegasus	DDO 216	Im V	23h28.6m, 14°45'	12.0	14.7	4'.6 by 3'.0	1700	Peg

1978. The 13-inch showed a 4' diameter uniform haze which was visible about 50 percent of the time. Psi^2 Piscium must be kept out of the field of view to have any chance of seeing the galaxy. **The Aquarius Dwarf** (DDO 210) is extremely difficult, showing a round 2' glow which disappeared from view at any powers over 118x. A 10th-magnitude star is situated 5' north. Save **IC 5152** for a night at a low-latitude site. At a declination of -51°, it skims the creosote bushes even from southern Arizona sites. Surprisingly, it was quite easy, showing a 3' by 2' glow with an 8th-magnitude star embedded in its western end. With several attempts racked up, I had no success in seeing **SagDIG**, the Sagittarius Dwarf Irregular Galaxy with the 13-inch. However, Houston amateur Larry Mitchell and I were able to pick this one out at the 1991 Texas Star Party. His 24-inch telescope resolved about two dozen faint stars over a region 2' across. It should appear about 3' by 2', and may be viewable from a Southern Hemisphere site.

Yes, there are Local Group galaxies in the Spring sky. But with objects appearing in both Aquarius and Virgo, the prospects of any "Local Group marathons" are pretty bleak! **Leo A** is another very distant member of the group, at over 1500 kpc. In the 13-inch, I was able to detect a faint, round glow about 4' across surrounding five 14th-magnitude field stars. **GR 8** was originally thought to be a member of the Virgo Cluster, and was catalogued as such by Gibson Reaves (1956). Eleven years elapsed before its Local Group membership was confirmed (Hodge, 1967). It is one of the smallest irregular galaxies known, at 50 parsecs across, the size of an ordinary globular cluster. GR 8 is situated less than 10' west of the bright edge-on galaxy NGC 4866 which is a distraction at any magnification. The 13-inch working at 215x showed a 14th-magnitude star at its position with a very small glow extending northward from this star measuring maybe 1' by 0.5' at best. GR 8 was clearly visible in Larry Mitchell's 24-inch scope as a round glow, 2' across, with some central brightening. Closing out the irregulars, we find **UGC-A86.** This one was imaged in 1990 with the 5-meter Hale telescope. Based on photometry and the degree to which it was resolved, the distance is tentatively set around 600 kpc, well within the Local Group envelope. Several futile attempts with the 13-inch showed nothing at its position.

The Ellipticals

Until recently, the dwarf ellipticals of the Local Group had been referred to as "The Seven Dwarfs." The addition of two more in 1990 brings the total to nine. As these are nearby, a necessary condition for them to be seen at all, they typically appear quite large and of incredibly low surface brightness. Since many of these have a roughly spherical shape, they resemble globular clusters. But they are not simply glorified globulars, as they show a younger population of stars. The largest dwarf elliptical even has globulars of its own!

Harlow Shapley found a smudge of light on a Harvard 24-inch refractor plate in 1937. In order to confirm this object, he had to go back 30 years to a photograph taken with a 3-inch aperture camera. They obviously didn't have hypered Konica 1600 back then — the image was a 72 hour exposure! The object was referred to as the **Sculptor System.** This would be the first discovery of a Local Group dwarf elliptical. The Sculptor System covers an immense area of the sky being over 1° in diameter. The best one can hope to see is a

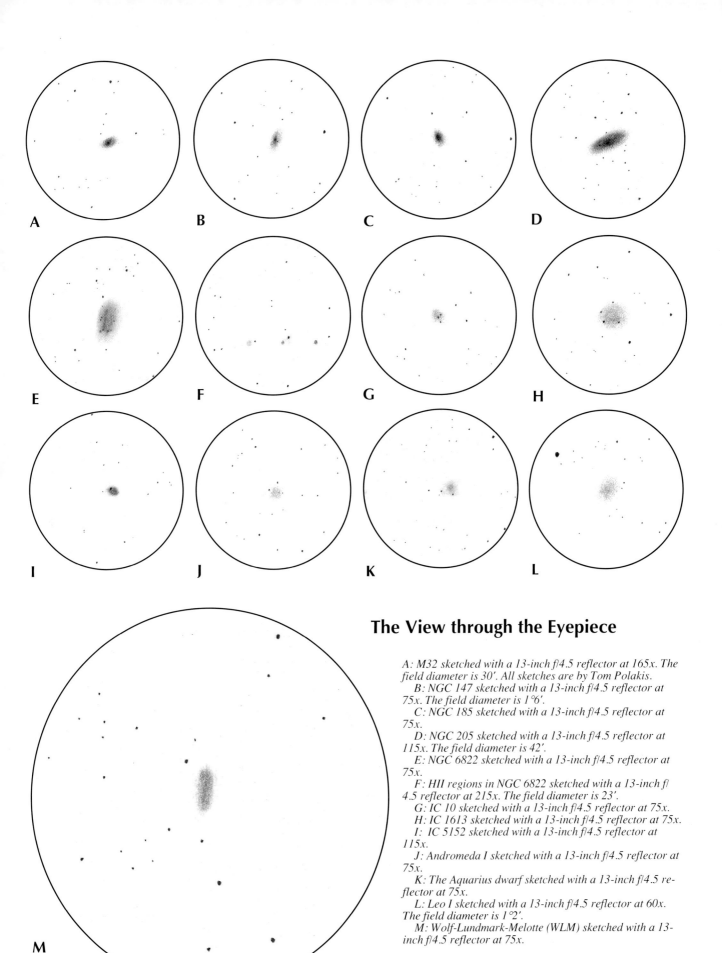

The View through the Eyepiece

A: M32 sketched with a 13-inch f/4.5 reflector at 165x. The field diameter is 30'. All sketches are by Tom Polakis.
  B: NGC 147 sketched with a 13-inch f/4.5 reflector at 75x. The field diameter is 1°6'.
  C: NGC 185 sketched with a 13-inch f/4.5 reflector at 75x.
  D: NGC 205 sketched with a 13-inch f/4.5 reflector at 115x. The field diameter is 42'.
  E: NGC 6822 sketched with a 13-inch f/4.5 reflector at 75x.
  F: HII regions in NGC 6822 sketched with a 13-inch f/4.5 reflector at 215x. The field diameter is 23'.
  G: IC 10 sketched with a 13-inch f/4.5 reflector at 75x.
  H: IC 1613 sketched with a 13-inch f/4.5 reflector at 75x.
  I: IC 5152 sketched with a 13-inch f/4.5 reflector at 115x.
  J: Andromeda I sketched with a 13-inch f/4.5 reflector at 75x.
  K: The Aquarius dwarf sketched with a 13-inch f/4.5 reflector at 75x.
  L: Leo I sketched with a 13-inch f/4.5 reflector at 60x. The field diameter is 1°2'.
  M: Wolf-Lundmark-Melotte (WLM) sketched with a 13-inch f/4.5 reflector at 75x.

FURTHER READING

Baade, W. 1950, *Publ. Univ. Michigan Obs.*, 10, 10.
Higgins, D. 1990, *Deep Sky* #32, 24.
Hodge, P. 1961, *AJ*, 66, 83.
—. 1967, *ApJ*, 148, 719.
—. 1974, *PASP*, 86, 289.
—. 1977, *ApJS*, 33, 69.
Hodge, P., Lee, M.G., and Gurwell, M. 1990 (pre-publication)
Hubble, E. 1925, *ApJ*, 62, 409.
Irwin, M.J. et al. 1990, *MNRAS*, 244, 16P.
Lavery, R.J. 1990 November 29, *IAU Circular No. 5139*.
Melotte, P.J. 1926, *MNRAS*, 86, 636.
Saha, A., and Hoessel, J. 1991, *AJ*, 101, 465.
Reaves, G. 1956, *AJ*, 61, 69.
Skiff, B.A. 1983, *Deep Sky* #4, 18.
—. 1984, *Deep Sky* #8, 8.
van den Bergh, S. 1972, *ApJ*, 171, L31.
Yahil, A., Tammann, G.A., and Sandage, A. 1977, *ApJ*, 217, 903.

AJ = The Astronomical Journal
ApJ = The Astrophysical Journal
ApJS = The Astrophysical Journal Supplement Series
MNRAS = Monthly Notices of the Royal Astronomical Society
PASP = Publications of the Astronomical Society of the Pacific

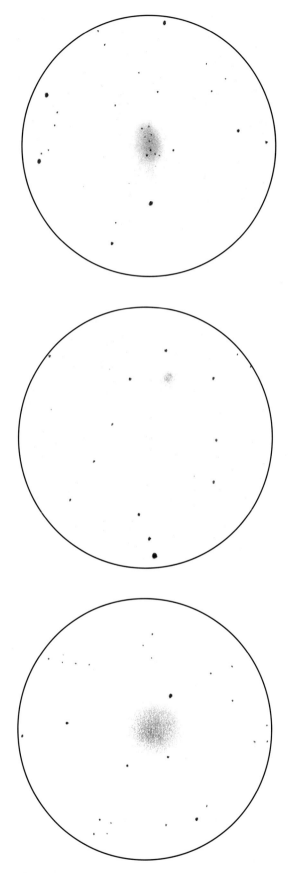

Top: The Pegasus dwarf sketched with a 13-inch f/4.5 reflector at 75x. Center: The SagDIG field sketched with a 13-inch f/4.5 reflector at 310x. The field diameter is 15'. Bottom: The Ursa Minor dwarf sketched with a 13-inch f/4.5 reflector at 75x. All sketches are by Tom Polakis.

subtle glow just above the skyglow. After several failed attempts with my 13-inch, I tried using observing buddy Rich Walker's 8-inch f/5 with a 20mm eyepiece giving a field of view over 1.5° across. By rocking the tube, a sky brightening could just be seen over about a 1/2° diameter region. Similarly, Steve Coe saw it best with a 4 1/4-inch f/4 rich field scope and a 32mm eyepiece. Again, Southern Hemisphere observers are challenged to see this as it passes near their zenith.

Very soon after discovering "the system in Sculptor," the **Fornax System** was found. This turned out to be the most massive of the dwarf ellipticals, being home to five (maybe six) globular clusters (Hodge, 1961). The Fornax System requires similar observing techniques as used for the Sculptor System. Rich Walker's 8-inch again did the trick in showing it best, although the 13-inch showed traces at 75x. The glow is 20' in diameter and round with no brightening toward the center. A 9th-magnitude star, plotted on *Uranometria 2000.0*, is centered in the glow. Using this "central star" as a starting point, four of the five globular clusters can be observed, even if the galaxy itself cannot. Located north of the central star, No. 3 is bright enough to warrant a listing in the *New General Catalogue* as NGC 1049. It looks similar to a planetary nebula at 310x, being 20" in diameter with a bright center. No. 4 is one magnitude fainter than NGC 1049 and appears 10" in diameter with a pretty bright center. No. 2 is faint and large, 20" in diameter but of low surface brightness. Finally, globular No. 5 appears about 15" in diameter with some slight central brightening. Only globular No. 1 was too faint for the 13-inch.

Swinging far to the north shows a pair of large, similar gal-

Left: Because of its proximity to Regulus and low surface brightness, Leo I is an extremely challenging galaxy for visual observers. Photo by Chris Schur (12.5-inch f/5 reflector, hypersensitized Tech Pan film, 60-minute exposure).

Below left: Globular clusters in the Fornax galaxy cluster. Photo courtesy Paul Hodge.

Below: IC 10 in Cassiopeia. Photo by Martin C. Germano (8-inch f/5 reflector, hypersensitized Tech Pan, 75-minute exposure).

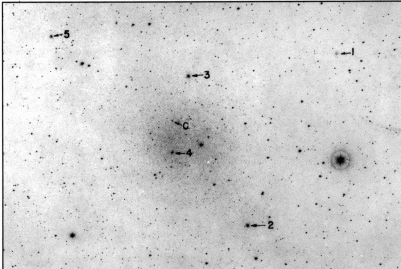

axies, the **Draco Dwarf** and the **Ursa Minor Dwarf.** These were found in the 1950s on the *POSS* plates. They are close but small, and suffer from the same low surface brightness as Shapley's original two dwarf ellipticals. The Draco Dwarf shows a size of 20' by 15', elongated east-west at 75x. It is completely uniform and very elusive visually. A faint grouping of 20 stars appears on the southern edge of the galaxy's glow. The Ursa Minor Dwarf is slightly brighter than the Draco Dwarf. It appears round and only showed its central 5' in my scope. This must simply be the brightest portion of a huge galaxy. Four 13th-magnitude stars cover the face of the galaxy. Scattered light from two bright stars to the northeast must be kept out of the field.

Leo I, also known as the Regulus System, is one of the most challenging Local Group galaxies to photograph. Many attempts were made with the 200-inch telescope, including some in which Walter Baade (1950) had baffles installed to try to combat the glare of a 1st-magnitude star just 20' away. I found Leo I to be pretty easy, elongated east-to-west 8' by 6' with a uniform inner 6' by 4', then tapering off gradually outside of this. A narrow apparent field eyepiece fares better than a sophisticated wide-field design in eliminating glare from Regulus. **Leo II** is another distant Milky Way companion at roughly the same distance as Leo I, to its southwest. It appeared at low magnification as a brightening of the field measuring 15' across, similar in surface brightness to the Sculptor System. **The Sextans Dwarf** was discovered early in 1990 on UK Schmidt plates (Kwin *et al.*, 1990). This galaxy is considered to be at the limit of what can be detected on these plates, even though the resolved image extends to over 1 1/2°. I saw no sign of it in the 13-inch as it passed near the meridian.

Two new dwarf ellipticals have joined the club recently. **The Carina Dwarf** was found on UK Schmidt plates in 1975. It is much less populous but otherwise quite similar to the Sculptor and Fornax dwarfs. Brian Skiff saw nothing obvious at its position with his 6-inch refractor working at 50x while in Chile. Finally, as I was putting this article together, a new Local Group galaxy, the **Tucana Dwarf**, was announced on an IAU Circular card (Lavery, 1990). Its color-magnitude diagram indicates that at 290 kpc, it is well within the confines of the Local Group. Both the Carina Dwarf and Tucana Dwarf are too low to observe from the northern hemisphere, but *Deep Sky* welcomes observations from southern hemisphere readers.

A tour through the Local Group gave me more of what I try to grasp through amateur astronomy, a perspective on our locale in the universe. Here is the word "local" used in its broadest sense, and yet these objects do seem somehow close. By observing these extremely faint objects, we get more clues about the scale of the universe. □

Tom Polakis is a contributing editor of Deep Sky *who observes from dark sites near his home in Tempe, Arizona.*

Observing the Morphology of Galaxies

by Alan Goldstein

An 8-inch scope shows the broad bar and faint outer arms of NGC 7479 in Pegasus, a textbook SBb galaxy. Photo by Kim Zussman (11-inch f/10 SCT, hypered Tech Pan film, 150-minute exposure).

Every reader of this magazine knows that galaxies vary over a wide spectrum of types. This fact was known before anyone knew what a galaxy really was! Observers like Sir William and John Herschel and Lord Rosse observed spiral patterns and other more subtle hints of structure with their telescopes more than one hundred years ago. You can do the same today.

The large apertures of many amateur telescopes make observing detail in galaxies relatively easy. Naturally, the closer the galaxy, the greater the details that can be resolved. This "detail" is often related to the classification to which the galaxy in question belongs. For example, spiral structure is indicative of the classification of normal or barred spirals (or occasionally the transitional barred spiral which falls between the two). Such detail is visible in galaxies that are relatively close to the Milky Way. Observing detail in galaxies, and understanding how that detail represents a particular type of galaxy, makes a wonderful observing project.

Two galaxies in the autumn sky reveal spiral structure and H II regions even with small instruments: M31 and M33. However, these galaxies have been discussed in great detail in earlier issues of this magazine. Consequently, we'll take a look at other galaxies that may be overlooked by observers who have not developed a comprehensive observing program.

As you read over this article and take your copy of the magazine outside this autumn, try to observe examples from each of the classes of galaxies we'll discuss. Observers with large aperture telescopes and keen eyes will be able to see structural detail that give clues to a galaxy's classification. Details to note include the proportion of the arms to the diameter of the central hub in spiral galaxies and the degree of oblateness in elliptical galaxies. Observers with small scopes will have a more challenging time spotting these details, but they are certainly visible under superior conditions.

Before we investigate individual galaxies, let's review the classification scheme. Edwin Hubble developed the basic system used to classify galaxies that is used today. He broke down the classification of normal spirals as Sa, Sb, Sc, and the barred spirals as SBa, SBb, and SBc.(Sa and SBa types had the largest central hubs and the least defined spiral arms.)

Elliptical galaxies were classified from E0, if circular in shape, to E7, if highly elliptical. Irregular galaxies were those with no organized structure. Peculiar galaxies were those that "fell between the cracks" in the classification scheme. Hubble soon realized that there were galaxies that fell between the spiral and elliptical classification in the "evolutionary sense" (there is no evidence to believe that galaxies evolve through the classification system). He designated these galaxies as type S0/SB0 and called them lenticular (lens-shaped) galaxies.

More recently other modifications have been made. Spirals with a distinct, though poorly developed, bar-shaped hub have been designated as transitional barred spirals — Sb type. Spirals with poorly organized arm structures have been designated Sd (an extension from Sc) and Sm (magellanic-type, with very poorly developed arms, like the Large Magellanic Cloud). In addition, the differences between the major types of galaxies have been smoothed out by introducing transitions between them (i.e. Sab, Sbc, Scd, etc.). Sidney van den Bergh introduced an "anemic" spiral galaxy which is characterized by a low surface brightness, a lot of interstellar gas, and a low concentration of stars.

The best examples of the Sa-type spiral galaxy are **NGC 7814** in Pegasus and **NGC 681** in Cetus. Both are in the 12th-magnitude range, although NGC 7814 is slightly brighter. In addition, both are highly inclined to our line of sight and appear nearly edge-on. NGC 7814 has a very thin equatorial dust lane, which may be glimpsed with a large aperture telescope. NGC 681 is a fainter version of the Sombrero galaxy, M104. Do you see its dark lane? What is the smallest aperture necessary to observe the shape? Unlike Sb and Sc galaxies, edge-on Sa-type spirals are still rather rotund in appearance because of their huge central hubs.

There are many choice Sb-type galaxies gracing the autumn sky. Yet each one is unique. Two examples are **NGC 7217** and **NGC 7331** in Pegasus. (NGC 7331 is the brightest galaxy in the constellation.) If you have tried to locate Stephan's Quintet, chances are NGC 7331 was your starting point. NGC 7217 is less well-known. At 11th magnitude it is a bright target for most telescopes. Picking up arm structure is certainly a challenge. This galaxy's arms are very fine and have a low surface brightness when compared to the central hub.

NGC 891 in Andromeda is another well-known Sb galaxy. After NGC 4565 in Coma Berenices, this galaxy is the best galaxy in which to observe an equatorial dust lane. Its surface brightness is fairly low, so dark skies are required for a good view. The view is one you are not likely to forget!

NGC 772 falls between the Sb and Sc classification. It is the brightest galaxy in Aries. This galaxy has a companion, NGC 770 (described below), and is experiencing gravitational distortions in the arms due to the interaction. A large aperture telescope may reveal an oddly placed bar shape in the arms. Can you find it?

There are a plethora of Sc spiral galaxies in the evening sky. Besides M33, **NGC 253** in Sculptor is one of the best. It does not take a lot of aperture to see detail, generally in the form of light and dark mottling. The spiral arms are very difficult to resolve, even in photographs, because this galaxy is highly inclined to our line of sight. With a large telescope, the galaxy is one of breathtaking beauty. I've detected mottling in NGC 253 with a 1-inch refractor under the most ideal sky conditions. What can you see?

NGC 6946 in Cepheus is another nice Sc galaxy. Its position in the rich star fields of the Milky Way and its nearly face-on appearance add to the difficulty in seeing it. You should pick it up without difficulty with a 6-inch telescope. In addition to a low surface brightness, this face-on galaxy lies behind galactic dust, which dims the galaxy by about two magnitudes. Its spiral arms are visible with large aperture instruments, but can you see the largest H II region? I have not yet been able to detect it.

Sd spiral galaxies tend to be on the low-mass side. Perhaps it is this low mass that makes spiral structure so loose and hard to define. The two brightest galaxies of this class have southern declinations. While both galaxies have relatively high photographic magnitudes (around 9.5), they have low surface brightnesses and are deceptively dim. With today's large aperture scopes, both offer a challenge to resolve arm structure. **NGC 7793** in Sculptor has more ill-defined arms. **NGC 247** in Cetus has one relatively conspicuous arm. What can you see?

One of the most difficult targets for observers is **IC 10**, a very loose spiral galaxy in Cassiopeia. Like NGC 6946 described above, it is obscured by interstellar dust and has a low surface brightness. The spiral designation may be considered arguable. Sdm types, like Sm galaxies, have very poorly organized arms. The structure is only visible with long-exposure photographs. This writer is confident in saying you will not see any structure. The challenge for you is observing this "ghost" galaxy.

NGC 6951 is another less well-known galaxy in Cepheus. At almost 12th magnitude, it can be observed in most telescopes. This writer is not familiar with amateurs observing any structural detail in this transitional barred spiral galaxy. Detail typically appears as an oval glow with a brighter nucleus.

The only bright SBa galaxy in the autumn sky, **NGC 7410**, is located in the far southerly constellation of Grus. Those with access to a -39° declination should be able to pick up this bright galaxy even in small scopes. Arm development in the SBa type is weak at best. Observers in southerly latitudes may be able to pick out structural detail in NGC 7410.

Loose, knotty arms characterize Sculptor's low surface brightness galaxy NGC 247, an Sd galaxy. Photo by Martin C. Germano (8-inch f/5 reflector, hypered Tech Pan film, 45-minute exposure).

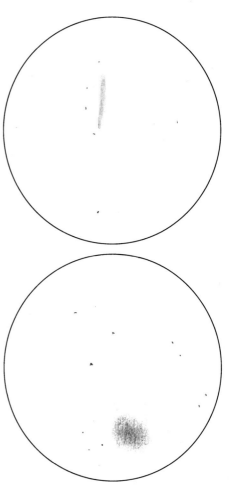

Left: Sculptor's NGC 55 is an SBbc galaxy that shows a bright nucleus and a single bright patch in its arms in small telescopes. Photo by Martin C. Germano (8-inch f/10 SCT, 103a-F film, 55-minute exposure).
Top: Visually, an 8-inch f/10 SCT shows NGC 55 as a slender streak of light surrounded by a chain of faint stars. Sketch by Alan Goldstein (8-inch f/10 SCT at 50x).
Above: The Sc galaxy NGC 6946 appears as a round patch of low surface brightness light. Sketch by Alan Goldstein (12-inch f/4 reflector at 35x).

NGC 7479 is a beautiful barred spiral (SBb) in Pegasus. At 11.6 magnitude, it can be seen with small telescopes. Larger instruments are needed for resolving the arms. In most scopes this object is little more than an elongated glow. **NGC 925** in Triangulum is about a magnitude brighter than NGC 7479. In large instruments an arm-shaped pattern can be seen in the area south following the nucleus. Its bar-shaped hub is distinct from the glow of the disk.

NGC 55 in Sculptor is very bright and would have no doubt been seen by Messier if it was higher in the sky. At 7.8 magnitude this SBbc galaxy can be seen in binoculars. An 8-inch telescope can clearly resolve two bright spots. The brightest spot is in the hub, the other is a star cloud forming the junction with the arm. The orientation of the galaxy makes details scanty at best. The faint glow trailing away from the hub is another bar-shaped hub that extends away from us in space. This galaxy is being viewed from a unique perspective!

Three SBc barred spirals in autumn sky are all interesting and not at all alike. **NGC 7640** in Andromeda glows at about magnitude 10.5. Its classification was very difficult to determine because of its edge-on orientation. Unlike NGC 55, this galaxy's arm structure could be seen through photographic rectification (described in detail in *Deep Sky* #12 [Fall 1985]). Like NGC 925, one arm is apparent with large aperture telescopes in the northeastern direction.

NGC 672 in Triangulum is a loosely structured SBc galaxy. This may be due, in part, to a possible interaction with IC 1727, an even looser barred spiral nearby. The bar and subtle indications of an arm may be seen with large optics. With smaller scopes, NGC 672 has a distinctly elongated appearance. This galaxy is highly inclined to our line of sight; however the arms are open and still readily visible in photographs.

NGC 1097 in Fornax is another bright galaxy. Again, its southerly declination keeps it off the "top ten list." In photographs, two large, well-developed arms are visible, beginning at each end of the bar-shaped hub. Halton C. Arp noted that both arms appeared to be sheared by jets emanating from the nucleus. This was in one of the few spiral galaxies known with exploding jets. The arms can be visible with moderate-sized telescopes, and a stellar nucleus can also be seen. It is unlikely that the kinks in the arms, where the shearing occurs, will be seen visually.

NGC 7457 is a type S0a, a lenticular galaxy, found in Pegasus. This galaxy is small, at 12.2 magnitude, and requires a modest aperture to be seen. It is unlikely that any detail can be seen with a large aperture scope.

NGC 404 is another S0a-type galaxy. Located in the glare of Beta Andromedae, it can be seen with small instruments. It appears as a round glow slightly

Kim Zussman's photograph of NGC 6946 shows a tiny nucleus surrounded by loose, dim arms set in a rich star field. Zussman used an 11-inch f/10 SCT, hyped Tech Pan and a 120-minute exposure.

Another bright Sc galaxy is M33 in Triangulum, a naked-eye object under a very dark sky. Telescopes show M33's nearly face-on arms and patchy distribuition of light. Photo by Martin C. Germano (8-inch f/5 reflector, hypered Tech Pan film, 45-minute exposure)

Galaxies by Morphological Type

Sa: NGC 681, NGC 7814
Sb: NGC 488, NGC 891, NGC 7217, NGC 7331, NGC 7723
Sbc: NGC 772, NGC 3344. NGC 7469
Sc: M33, M101, NGC 253, NGC 300, NGC 2403, NGC 6946, NGC 7314, NGC 7448
Sd: NGC 247, NGC 4395, NGC 7793
Sdm: IC 10, IC 5152
Sm: IC 3356
Sb: NGC 6951
Sc: NGC 1232
SB: M81
SBa: NGC 7410
SBb: NGC 151, NGC 4236, NGC 7479
SBbc: NGC 55
SBc: NGC 672, NGC 925, NGC 1097, NGC 7640
SBm: NGC 2366
S0p: NGC 7625, NGC 7679
S0a: NGC 404, NGC 474, NGC 7457.
S0ap: NGC 128
S0b: NGC 1201
SB0: NGC 16, NGC 1023
Aa: NGC 718
E0: NGC 750/1
E0p: NGC 7742
E3: NGC 7619
E4: IC 1459
E5: NGC 770
E6: NGC 147
I: IC 1613
IBm: NGC 1156
Ring: NGC 985, NGC 7714/15
Sp? or Ep?: NGC 1275

brightening toward the middle. NGC 404 bears magnification well, which allows it to be viewed away from Beta.

Located in Pisces, **NGC 474** is an S0a that interacts with **NGC 470**, an Sc galaxy. The pair is 227 in Halton C. Arp's *Atlas of Peculiar Galaxies*. NGC 474 appears as a tiny circular glow, considerably smaller than the fainter NGC 470. The peculiar halo surrounding NGC 474 is much too faint to be seen visually.

NGC 128 in Pisces is another peculiar S0a-type lenticular galaxy. The curious feature with this galaxy is the box-shaped central hub. Its edge-on appearance should be readily visible with larger aperture. The odd shape of the hub may be visible with the largest optics. Companions **NGC 127** and **NGC 130** may be seen nearby, as well as some other galaxies.

Located in Fornax, **NGC 1201** is not the brightest galaxy in the constellation, but it has the distinction of being the best example of the classification S0b in autumn. This S0b appears as an elongated glow. The brighter nuclear region is distinct in a larger telescope. The southerly declination makes it a greater challenge for northern observers, but it is bright enough to be seen clearly with a 6-inch aperture.

NGC 16 is faint but is an example of one of the uncommon SB0 class (barred lenticular galaxy). **NGC 1023** in Perseus is another SB0-type and is considerably brighter. Both appear as elongate glows, NGC 1023 more so. Neither galaxy shows any detail with larger optics.

NGC 7625 is a peculiar lenticular galaxy in Pegasus. Also designated Arp 212, this galaxy is very faint, appears as a roundish wisp, and is considered a challenge to observe in smaller telescopes. **NGC 7679** is another faint peculiar galaxy. I have seen this galaxy, also known as Arp 216, described as a peculiar lenticular galaxy or a giant intergalactic H II region. Whatever the case, this weird object is located in Pisces and is worth tracking down. NGC 7682 lies in close proximity.

NGC 718 has been classified in the past as a Sa or Sb spiral. It appears to meet the criteria for an anemic spiral galaxy designated type Aa. Located in Pisces, this photographic magnitude-12.5 galaxy is the brightest anemic galaxy in the autumn sky. It appears as a soft, roundish glow.

There are several noteworthy galaxies of type E0. **NGC 7742** in Pegasus is listed as having peculiar characteristics. At photographic magnitude 12.2, this round object can be seen in modest instruments. Typical of elliptical galaxies, there are no "details" to be seen. **NGC 750** and **NGC 751** are a faint pair of E0 galaxies in Triangulum. Both galaxies glow at 13th magnitude and are in contact. An observer who is not careful may note this pair as one elongated system! This error is made in the *Skalnate Pleso Atlas Catalogue*. This catalogue lists the dimensions as 0.6' by 0.3'.

The odd S0a galaxy NGC 404 is difficult to observe because it lies adjacent to the dazzling star Beta Andromedae. Photo by Martin C. Germano (8-inch f/5 reflector, hypered Tech Pan film 35-minute exposure).

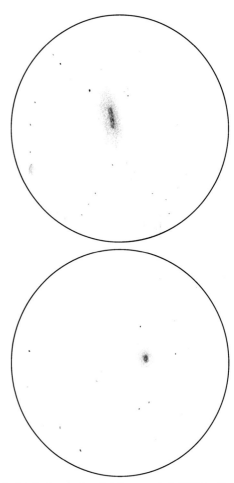

Left: The low surface brightness spiral M101 in Ursa Major is a tremendous example of an Sc-spiral seen face on. Photo by Kim Zussman (14.5-inch f/8 Cassegrain, hypersensitized Tech Pan film, 100-minute exposure).
Top: NGC 672 is an SBc galaxy that appears as a bright, elongated disk surrounded by a faint glow. Sketch by Alan Goldstein (11-inch f/10 SCT at 175x). The satellite galaxy IC 1727 lies at the edge of the field.
Above: A typical Sb galaxy, NGC 7723 is a roundish smear of light in small telescopes. Sketch by Alan Goldstein (8-inch f/10 SCT at 50x).

However, each galaxy measures 0.3' by 0.3'. A large telescope is useful for a clear look at this pair, which is also known as Arp 166.

Located in Pisces, **NGC 7619** is little more than a moderately bright E3 galaxy. This elliptical galaxy shows a brightening toward the center and tends to be slightly elongated. The best example of an E4 galaxy in the autumn sky is IC 1459 in the southern constellation Grus. If sky conditions permit, it appears as an oval patch. Observers with better southern horizons will see the brighter central glow easier. **NGC 770** is a small E5 galaxy next to NGC 772 in Aries. It is faint, about 13th magnitude, and required a moderate aperture and good skies to be seen well.

NGC 147 in Cassiopeia is a dwarf E6 galaxy. It is somewhat ghostly in appearance. NGC 147 is a member of the Local Group and is about the same distance as M31. The central condensation is not apparent as with other elliptical galaxies. **IC 1613** is a very faint irregular galaxy in Cetus. It is a challenge to resolve and requires a large aperture and short focal length. At 11' across, it is about the same diameter as the long axis of NGC 205.

NGC 1156 is a fairly bright IBm. It is a magellanic-type barred irregular galaxy in the constellation Aries. At photographic magnitude 11.6, it can be seen in a 6-inch telescope. This galaxy doesn't reveal much detail with larger scopes but is definitely worth the effort to observe.

There are several other galaxies worth observing, because they do not fall into the regular classification scheme. These peculiar galaxies are noteworthy for different reasons. **NGC 7714/5** is a ring galaxy, an effect caused by intergalactic collisions. All ring galaxies have a low surface brightness, so don't plan on seeing a ring shape!

NGC 985 in Cetus is an example of the unlikely case of both a ring galaxy and a Seyfert galaxy with an exploding nucleus. Like NGC 7714/5, don't plan to see much detail, although this one has a bright stellar nucleus (brighter than the typical magnitude 13.5 galaxy).

NGC 1275 is a peculiar Seyfert galaxy and a bright radio emission source. It is a small member of the Perseus galaxy cluster, but it is not difficult to observe with a 6-inch aperture. The filamentary structure is much too faint for visual observation.

Get outside this autumn and see how many different classes of galaxies you can see. The galaxies I have mentioned represent just the tip of the iceberg. Remember that to see the maximum amount of detail possible with your instrument the skies must be optimum. Naturally, *Deep Sky* magazine and the National Deep Sky Observers Society are always interested in receiving copies of your sketches, observing notes, photographs, or other records that pertain to observing sttructure in galaxies.

Observing Interacting GALAXIES

by Alan Goldstein

Imagine standing out beneath the sparkling spring sky, ready for another warm observing season. You are scanning near the handle of the Big Dipper, and you come across a pair of fuzzy patches almost in contact. You just glimpse a hint of arm structure in the larger patch, and that one arm appears to touch the second misty spot.

This scene is familiar to all observers who have ever looked at the bright and well-known galaxy M51 in Canes Venatici, the most famous of the interacting galaxies. M51 is better known by its nickname the "Whirlpool Galaxy." When viewed with a large telescope under a dark sky, the Whirlpool is one of the most stunning of all deep-sky objects. However, most observers may be a surprised that fifteen other interacting systems lie within a 20° radius of M51 and are visible in backyard telescopes.

What is an interacting galaxy? It is a galaxy influenced significantly by the gravitational pull of another galaxy. This gravitational attraction may be rather one-sided, when a large galaxy interacts with a small galaxy, for instance, or it can be an equal relationship between two similar galactic systems.

Every type of galaxy is vulnerable to interaction with others that pass nearby. The most common pairs of interacting galaxies seem to be ellipticals with ellipticals and ellipticals with spirals, followed by spirals with spirals, spirals with irregulars, ellipticals with irregulars, and irregulars with irregulars. Even the two best-known galaxies, the Milky Way and the Andromeda Galaxy, are both interacting: The Milky Way with both the Large and Small Magellanic Clouds, and the Andromeda Galaxy with M32.

Interacting galaxies are perhaps most common in rich, dense clusters of galaxies. Some groups of galaxies, like the Hercules, Coma Berenices, and Coma/Virgo clusters, each have a dozen or more interacting members. Some theorists in the evolution of galaxies propose that lenticular (S0) galaxies are created by frequently interacting spirals which cause a loss of gas. Galaxies have been found in various stages of lenticular-to-spiral transition, which supports this theory. And unusually high number of lenticulars — and low number of spirals — have been found near the centers of rich clusters, where interaction is likely to occur often. Other hypotheses include the contention that galaxy interactions help create the spiral arms in galaxies. This may explain why M51 has such a strongly-defined arm pattern, but it doesn't explain why many spirals are completely isolated.

Interacting galaxies produce an infinite number of results. Some of these, such as connecting clouds of neutral hydrogen, are observable only with radio telescopes. Other features include asymmetric arms, tails and bridges of matter, and odd diffuse haloes of gas. The most bizarre feature is produced when a small elliptical galaxy passes through a larger spiral galaxy, leaving a ring galaxy in its wake. (Unfortunately, there are no ring galaxies

M51 in Canes Venatici is the sky's brightest example of two galaxies clearly interacting. Photo by Kim Zussman (14.5-inch f/8 Cassegrain, hypersensitized Tech Pan film, 93-minute exposure).

Virgo's NGC 5363 (top) and NGC 5364 offer an unusual mixture of peculiar elliptical and peculiar spiral. Both are 10th-magnitude objects and are easily visible in small telescopes. Photo by K. Alexander Brownlee.

Opposite top: *Another interacting pair in Virgo is composed of two Sc-type spirals: NGC 5426 (bottom) is a magnitude 11.2 object, and NGC 5427 glows at magnitude 11.4. Photo by K. Alexander Brownlee.*

Opposite bottom: *The interacting pair containing NGC 4298 (left) and NGC 4302 lies in Coma Berenices. Both are Sc-type spirals. Photo by Chuck Vaughn (14-inch SCT at f/7, hypersensitized Tech Pan film, 60-minute exposure).*

brighter than magnitude 13.5)

Observing Techniques

Observing interacting galaxies can be a challenge, though it does not require more skills than observing run-of-the-mill faint galaxies does. The difficulty in observing interacting pairs really hinges on the brightness difference between galaxies, the shapes of the objects, and the separation between the objects.

Sketching interacting galaxies can be useful for recording your observations. Over a period of time it will allow you to compare different types of interacting galaxies and to maintain a record of observations you can refer back to. If you do not want to sketch deep-sky objects, at least write a short paragraph describing the appearances of the objects.

With a large telescope, 12-inches or more in aperture, you can observing some very challenging interacting galaxies. Try to observe connecting bridges, bright and dark patches near the point of contact, and other features. Large scopes also offer sheer numbers: with an 8-inch instrument you may observe some fifty interacting pairs, but with a 12.5-incher the list grows to 150.

Selected Galaxies

The following list is a guide to fourteen of the best interacting galaxy pairs, though it is not exhaustive even for small telescopes.

NGC 750 and **NGC 751**; designation = Arp 166, VV 189. These galaxies form an interacting pair in Triangulum, not far from the pinwheel galaxy M33. NGC 750 is a magnitude 12.2 elliptical measuring 1.6' by 1.3'; NGC 751 is a magnitude 12.5 peculiar elliptical measuring 1.3' by 1.3'. (All coordinates in this article are for epoch 2000.0.) This is a faint but interesting pair of ellipticals with outer envelopes of nebulosity in contact. You'll probably need a 12.5-inch scope to see this pair well.

NGC 3166 and **NGC 3169**. This pair in Sextans is composed of NGC 3166, a magnitude 10.6 barred spiral spanning 5.2' by 2.7', and NGC 3169, a magnitude 10.5 spiral measuring 4.8' by 3.2'. These galaxies are not visually in direct contact but lie close enough to each-other that they're considered interacting. NGC 3169 shows some apparent disturbances from the gravitational tug of NGC 3166. This pair is easy to spot in a 6-inch scope because it appears as a double elliptical patch of nebulosity.

NGC 3226 and **NGC 3227**; designation = Arp 94, VV 209. One of Leo's finest galaxy pairs consists of NGC 3226, a magnitude 11.4 elliptical some

2.8' by 2.5' across, and NGC 3227, a magnitude 10.8 spiral measuring 5.6' by 4.0'. NGC 3227 is a Seyfert galaxy, and its stellar nucleus is visible in large backyard scopes. In a 6-inch instrument this pair appears as two oval patches lying nearly in contact. An 8-inch scope shows a hint of nuclear brightening.

M105, **NGC 3384**, and **NGC 3389**. Leo's M105 trio contains M105, a magnitude 9.3 elliptical measuring 4.5' by 4.0', and two fainter galaxies: NGC 3384, a magnitude 10.0 elliptical measuring 5.9' by 2.6', and NGC 3389, a magnitude 11.8 spiral some 2.7' by 1.5' across. In photographs none of these galaxies appears very disturbed, with the possible exception of the spiral. A 6-inch scope shows all three galaxies, though the spiral requires an 8-incher to be seen well.

NGC 3395 and **NGC 3396**; designation = Arp 270, VV 246. Located in Leo Minor, this little-observed pair contains NGC 3395, a magnitude 12.1 spiral some 1.9' by 1.2' across, and NGC 3396, a magnitude 12.2 peculiar galaxy measuring 2.8' by 1.2'. While these two galaxies are both faint and small, they form one of the best examples of an interacting galaxy pair. Not only are both spirals in direct visual contact, but their mutual interaction causes both to become distorted. An 8-inch scope shows this pair as two galaxies in contact, aligned at right angles to each other.

NGC 4298 and **NGC 4302**. One of the best interacting pairs in Coma Berenices is formed by NGC 4298, a magnitude 11.4 spiral some 3.2' by 1.9' across, and NGC 4302, a magnitude 11.6 edge-on spiral measuring 5.2' by 1.1'. Using an 8-inch scope on a dark night, I only suspected NGC 4302. A 16-inch scope, on the other hand, will easily show 4302 and may even reveal its razor-thin dust lane.

NGC 4435 and **NGC 4438**; designation = Arp 120, VV 188. A splendid interacting pair in Virgo holds NGC 4435, a magnitude 10.9 elliptical some 3.0' by 1.9' across, and NGC 4438, a magnitude 10.1 spiral covering 9.3' by

At magnitude 9.3, the edge-on Sc-type spiral NGC 4631 is one of the finest galaxies in Canes Venatici. The galaxy's little companion, NGC 4627, however, may require a 10-inch or larger telescope for successful viewing. Photo by Kim Zussman (modified 11-inch f/10 SCT, hypersensitized Tech Pan film, 120-minute exposure).

Opposite top: *Also in Canes Venatici is the bizarre pair formed by NGC 4485 (top), a tiny irregular galaxy, and NGC 4490, a distorted Sc-type spiral. Photo by George Greaney.*

Opposite bottom: *Yet another pair in Canes Venatici holds NGC 5394 (bottom), a warped barred spiral, and NGC 5395, an Sb-type galaxy.*

3.9'. A member of the Virgo Cluster of galaxies, this pair is easy to spot with a small telescope. NGC 4438 is highly inclined but disturbed by the elliptical, so it is not very thin. Most backyard telescopes show, only the central portions of the spiral because the arms have a very low surface brightness.

NGC 4485 and **NGC 4490**; designation = 3C 272, Arp 269, VV 30. Canes Venatici contains a fabulous interacting pair containing NGC 4485, a magnitude 12.0 irregular spanning 2.4' by 1.7', and NGC 4490, a magnitude 9.8 spiral measuring 5.9' by 3.1'. This duo can be observed with a 4-inch scope under a dark sky. With an 8-inch instrument I have observed mottled brightening in

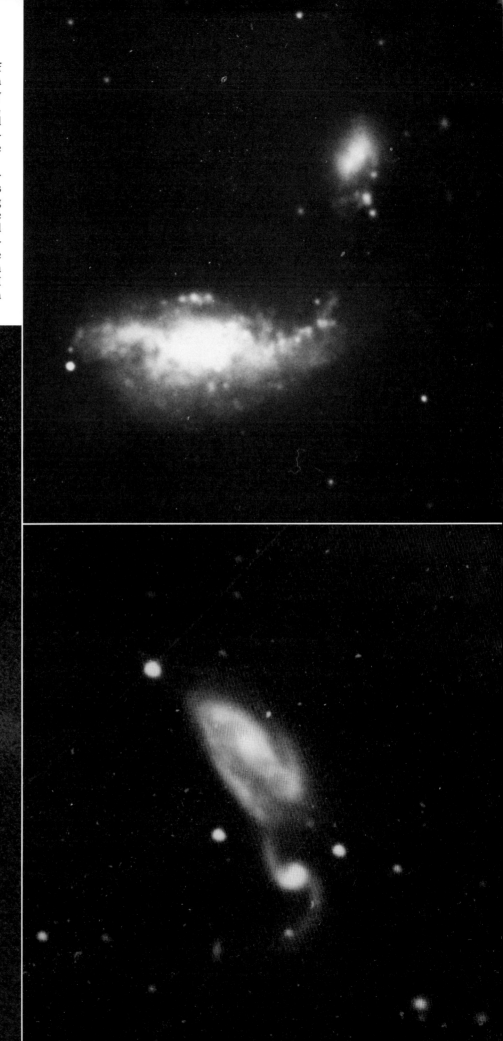

Three galaxies dominate this picture: the Sb-spiral NGC 5350 lies at top, while the interacting pair is made up of NGC 5353 (bottom) and the lenticular galaxy NGC 5354. Photo by Martin C. Germano (8-inch f/5 reflector, hypersensitized Tech Pan film, 65-minute exposure).

Opposite: *Leo Minor's contribution to the pool of interacting galaxies is NGC 3395 (top), an Sc-type spiral, and the peculiar galaxy NGC 3396.*

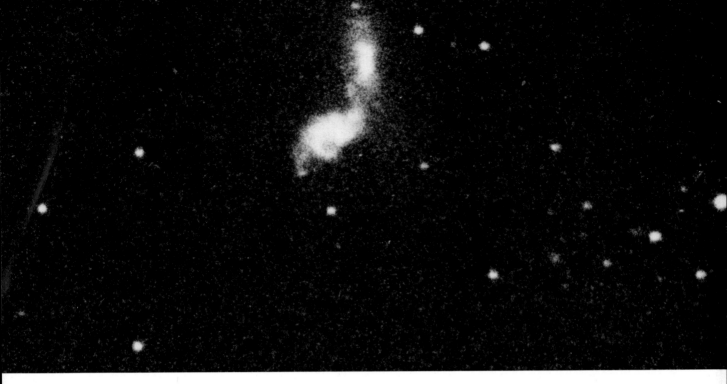

NGC 4490, which appears as a bar-shaped mass of gray nebulosity. NGC 4485 is a tiny patch of light without detail. The pair lies less than 1° northwest of the bright star Beta (β) Canum Venaticorum.

NGC 4627 and **NGC 4631**; designation = Arp 281. Also lying in Canes Venatici is the pair composed of NGC 4627, a magnitude 12.3 peculiar elliptical measuring 2.7' by 2.0', and NGC 4631, a magnitude 9.3 spiral covering 15.1' by 3.3'. The largest galaxy in Canes Venatici is NGC 4631, and it is visible as a bright, thin ray in any small telescope. Under a dark sky the dwarf elliptical companion can be glimpsed as a tiny fuzzy patch in a 6-inch scope.

M60 and **NGC 4647**; designation = Arp 116, VV 206. A fine interacting pair in Virgo contains M60, a magnitude 8.8 elliptical measuring 7.2' by 6.2', and NGC 4647, a magnitude 11.4 spiral measuring 3.0' by 2.5'. M60 is a giant elliptical galaxy in the Virgo Cluster and one of the brightest galaxies in Messier's catalogue. Messier missed the face-on spiral NGC 4647, a faint object that appears like a misty patch of light in a 6-inch scope.

NGC 4752 and **NGC 4762**. Virgo contains another fine interacting pair in NGC 4752, a magnitude 10.6 barred spiral measuring 4.7' by 2.6', and NGC 4762, a magnitude 10.2 barred spiral spanning 8.7' by 1.6'. NGC 4754 appears as an evenly-illuminated oval patch; 4' south of this object is 4762, one of the thinnest galaxies in the northern sky. In small telescopes its silvery needle of light is quite spectacular.

M51 and **NGC 5195**. The king of the interacting galaxies is M51 in Canes Venatici, one of the brightest galaxies in the sky. In small telescopes M51 appears as a double patch of nebulosity because a small companion galaxy, NGC-5195, is connected to M51 by one of the large spiral's arms. M51 is an Sc-type spiral glowing at magnitude 8.4 and measuring 11.0' by 7.8'. NGC-5195 is a magnitude 9.6 peculiar galaxy some 5.4' by 4.3' in extent. With a large backyard telescope the view of M51 is breathtaking: a 21-inch shows it with prominent spiral structure and a clearly visible bridge between the two galaxies.

NGC 5363 and **NGC 5364**. Virgo contains a spectacular interacting pair in NGC 5363, a magnitude 10.2 peculiar elliptical galaxy some 4.2' by 2.7' across, and NGC 5364, a magnitude 10.4 spiral measuring 7.1' by 5.0'. NGC 5363 appears as a bright round patch of nebulosity; the spiral is a bright, large, dish-shaped oval of light with a bright center and smooth halo of nebulosity.

NGC 5394 and **NGC 5395**; designation = Arp 84, VV 48. A challenging interacting galaxy is formed by the galaxies NGC 5394 and NGC 5395 in Canes Venatici. NGC 5394 is a type SB spiral, glowing at magnitude 11.6 and covering 3.1' by 1.7'. NGC 5395, on the other hand, is a magnitude 13.0 barred spiral measuring a mere 1.9' by 1.1'. This galaxy appears as a very small, fuzzy spot, while NGC 5394 looks like a much brighter, distinctly oval patch.

NGC 5426 and **NGC 5427**; designation = Arp 271, VV 21. Virgo contains another interesting interacting pair in NGC 5426, a magnitude 12.2 Sc-type spiral some 2.9' by 1.6' across, and NGC 5427, an Sc galaxy glowing at magnitude 11.4 and measuring 2.5' by 2.3'. In 6-inch telescope this pair appears as an elongated dim patch of nebulosity; larger telescopes hint at arm structure in these galaxies.

This list of interacting galaxies is not exhaustive, but it should provide you with some of the sky's best examples of these distant, changing extragalactic wanderers. As you observe these galaxies, keep in mind that the editors of *Deep Sky* look forward to hearing about your observations.

In 1976 Alan Goldstein founded the National Deep Sky Observers Society, the only nationwide group of active deep-sky observers in the United States. One year later he became a contributing editor of Deep Sky Monthly, *and he has contributed many articles to* ASTRONOMY *as well. This summer Alan will be married; the editors of* Deep Sky *wish him a very happy future.*

Observing the M81 Galaxy Group

by Tom Polakis

Like stars, galaxies are gravitationally bound to companions in the heavens. Much as our Milky Way Galaxy is a member of the Local Group, most galaxies that amateurs observe belong to larger-scale gatherings. Astronomers are beginning to see evidence of larger and larger scale structures, and in one of the large galaxy aggregations known as the Coma-Sculptor Cloud lies our own Local Group and its neighboring galaxy groups. Examples of neighboring galaxy groups appear in the constellations of Sculptor and Ursa Major. It is the later assemblage, also known as the M81 group, that we will discuss.

Local Group galaxies are found within a sphere 3 megaparsecs (Mpc) in diameter whose center lies between the Milky Way and M31. It is outside this sphere that we begin to see evidence of cosmological expansion in the spectra or galaxies. M81 group galaxies were the first outside of the Local Group in which the light curves of Cepheid variables in conjunction with redshirt data were used to home in on the elusive Hubble constant. These measurements have placed the distance to the center of the M81 group at 3.25 Mpc, or approximately 10 million light-years.

Membership in the M81 group is confirmed only when The distance to the suspected member is shown to be near that of M81. Distance estimates can be subject to interpretation, however, since many galaxies in this regime do not have bright Cepheid variable stars to use as standard candles. For this reason, there are about as many different M81 group membership lists as there are sources. Erik Holmberg (1950) first defined a list of fifteen M81 group members. This list has been revised many times since. Recently, the team of Karachentsev, Karachentsev, & Bîrngen (1985) photographed 37 additional dwarf galaxies near M81 with the Soviet 6-meter telescope, resolving many of them into stars on the plates. For our discussion, we'll focus on the list of fifteen definite members whose velocities have been measured, adopted from the *Nearby Galaxies Catalog* by R. Brent Tully (1987a). This catalog is intended lo be a companion to the masterful *Nearby Galaxies Atlas* (Tully & Fisher 1987), a multi-color collection of maps that gives an excellent three-dimensional representation of The large-scale structure of the nearby universe.

Complementing these two works is an article in the *Astrophysical Journal* by Tully (1987b) assigning the 2,367 galaxies plotted in the atlas to clouds, associations, and groups. Groups, such as the one found around M81, are simply small concentrations in larger features called associations, which in turn are part of still larger-scale structures called clouds. By performing an exhaustive analysis of their velocities, Tully was able to show that 64 percent of the galaxies in the sample lie in groups and only 1 percent of the sample are "at large" and not even associated with clouds.

For the visual observer, the M81 group presents a wide range of galaxies from star-party favorites to obscure dwarfs, barely visible on the *Palomar Observatory Sky Survey* (*POSS*) prints. A key to seeing the most detail in these objects, particularly in the faint dwarf galaxies, is to use the whole range of telescopic magnifications. The fainter galaxies often disappear from view at either too-high or too-low magnifications. For the lowest-surface-brightness objects, a tape recorder allows one to take notes without exposure to the red flashlight. A lesson about objects at high declinations: if you are using a large Newtonian on a German equatorial mount, be prepared to rotate the tube! Over a half-hour observation, the eyepiece height change on my 13-inch scope would result in beginning the observation crouching and finishing it standing on tip-toes.

Working our way from the brightest to faintest members, Let's take a tour of the M81 galaxy group. At the heart of the thirteen galaxies is none other than **M81** (NGC 3031). It is a classic example of an Sb spiral galaxy, with thin but well-defined spiral arms around a large nuclear region. Both in photographs and through the telescope, M81 resembles M31, the Andromeda Galaxy, another Sb spiral. M81 has proven to be an excellent nearby specimen for testing theories of spiral arm formation. Spiral structure in a galaxy cannot be explained by the same simple mechanism as that of cream stirred into your morning cup of coffee. Over time, the arms would wrap many times around the nucleus instead of just once or twice as is typically observed. The theory of density waves proposes that galactic spiral structure is a result of accumulation of material along the intersection or elliptical paths defining the orbit of stars, dust, and gas (Kaufman, 1987). Observations in the infrared and radio portion of the spectrum have supported the theory of density waves with the detection of HII regions in the arms and the smooth-to-ragged transition from the inner to outer portions of the arms.

In my 13-inch scope working at 75x, the initial impression of M81 is of a disk whose overall dimensions are 15' by 7' in position angle (p.a.) 150°. There is a very bright central glow surrounded by a core 5' by 3' in size whose brightness falls off linearly. The spiral arms are subtle but discernible. The

Galaxy M81 photographed by Bill and Sally Fletcher (16-inch f/4.5 reflector, hypersensitized Tech Pan film, taken on December 14, 1990, from Mt. Pinos, California.

eastern arm can be seen more easily as it bears off the south end of the galaxy. A couple of the brighter stellar associations can be seen in this arm at the south end of the galaxy and due east of its nucleus. The western arm has a much thicker and less welldefined appearance. A magnification of 165x showed the nucleus to be nonstellar. At 215x the appearance of the core region is similar to an elliptical galaxy elongated 2:1. The most striking difference from photographic images is differentiating the thick spiral arms from a small nuclear region, where photographs almost always burn this central area out.

Brian Skiff, working with a 6-inch f/8 refractor from Lowell Observatory's remote Anderson Mesa Site in northern Arizona, saw a very smooth texture to the galaxy at all powers, with no hint of concentric spiral structure. At 195x he saw the brightest region near The center to be nearly circular, 30" across, and with a strong sharp concentration to a completely stellar nucleus.

M82 (NGC 3034) is certainly one of the most-studied irregular galaxies in the heavens. Spectroscopic observations made in the 1960s at many wavelengths led astronomers to the conclusion that M82 is an exploding galaxy. The observations appeared to reveal gas and dust velocities that could only be associated with an explosion. But in the mid-1970s the bottom fell out of the explosion hypothesis with the failure to find a compact center to this explosion and the discovery that the inferred high gas speeds were due to "mirrors" of dust moving past M82 (Solinger, Morrison, & Markert, 1977). It now appears that M82 is a normal galaxy, albeit

Opposite page: M82 photographed by Martin C. Germano (8-inch f/10 SCT, hypered Tech Pan film, 60-minute exposure).

Top: NGC 2403 photographed by Harvey Freed (10-inch f/6 SCT, hypered Tech Pan film, 88-minute exposure).

Right: NGC 2336 photographed by Chuck Vaughn (14-inch SCT at f/7, hypered Tech Pan film, 60-minute exposure).

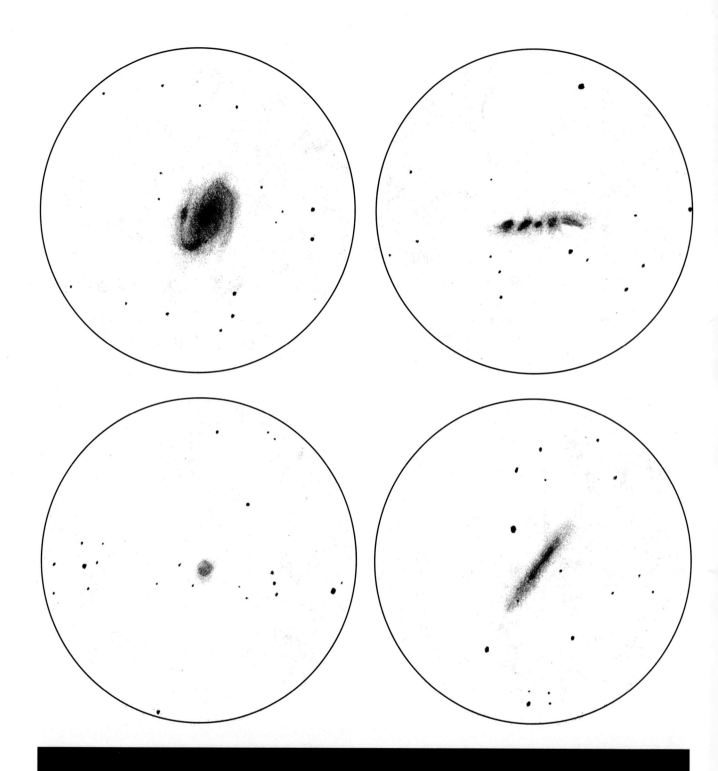

Clockwise, from top left: M81 sketched with a 13-inch f/4.5 reflector at 115x. The field diameter is 42'. The site was Sentinel, Arizona, and the date April 7, 1991. All sketches on this spread are by Tom Polakis.

Top right: The unusual galaxy M82 sketched with a 13-inch f/4.5 reflector at 215x, yielding a 24'-field.

Bottom left: The notoriously faint galaxy Holmberg I sketched with a 13-inch f/4.5 scope at 115x.

Bottom right: The galaxy NGC 4236 sketched with a 13-inch f/4.5 scope at 115x.

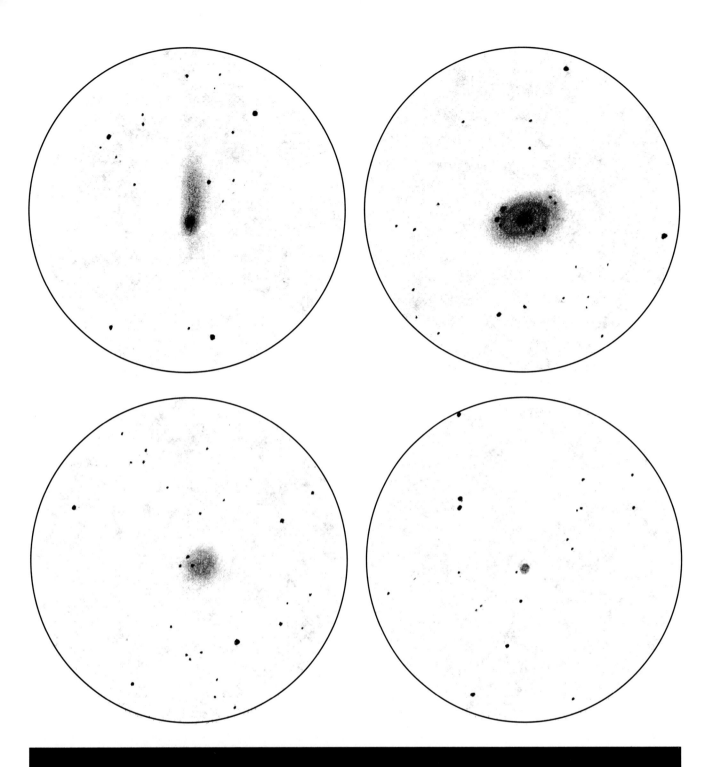

Clockwise, from top left: NGC 2366 sketched with a 13-inch f/4.5 scope at 215x.
Top right: NGC 2403 sketched with a 13-inch f/4.5 scope at 215x.
Bottom left: The elusive Holmberg II sketched with a 13-inch f/4.5 scope at 165x, yielding a 30' field.
Bottom right: The extremely challenging galaxy UGC 4459 sketched with a 13-inch f/4.5 scope at 115x.

Galaxies in the M81 Group

Name	Other	Type	R.A.(2000.0)Dec.	Mag.(V)	SB(V)	Size	Con.
NGC 2366	DDO 42	IB(s)m IV-V	7h28.9m +69°13'	10.9	14.3	8.0' by 3.5'	Cam
NGC 2403		SAB(s)cd	7h36.9m +65°36'	8.4	13.9	23.8' by 12.9'	Cam
Holmberg II	UGC 4305	Im IV-V	8h18.9m +70°43'	10.7	14.8	8.2' by 6.1'	UMa
M81 DW A	0818+71	IBm V	8h23.5m +71°03'	15.9	15.2	1.0' by 0.5'	UMa
UGC 4459	DDO 53	Im V	8h34.1m +66°10'	15.4(B)	16.6(B)	1.9' by 1.6'	UMa
Holmberg I	UGC 5139	IAB(s)m V	9h40.5m +71°11'	12.9	15.3	3.7' by 3.7'	UMa
NGC 2976		SAc pec	9h55.6m +67°55'	10.2	12.7	5.0' by 2.9'	UMa
M81	NGC 3031	SAB(s)ab I-II	9h55.6m +69°04'	6.7	13.1	22.1' by 12.6'	UMa
M82	NGC 3034	I0 sp	9h55.8m +69°41'	8.4	12.5	11.7' by 5.7'	UMa
NGC 3077		I0 pec	10h03.3m +68°44'	9.9	12.7	5.4' by 4.2'	UMa
M81 DW B	UGC 5423	Im IV	10h05.5m +70°22'	15.3(B)	15.8(B)	1.4' by 1.1'	UMa
IC 2574	UGC 5666	SAB(s)m IV-V	10h28.4m +68°25'	10.6	15.0	13.0' by 8.2'	UMa
UGC 6456		Pec	11h28.0m +78°59'	14.7(B)	15.3(B)	1.7' by 1.0'	UMa
NGC 4236		SB(s)dm IV	12h16.7m +69°28'	9.6	14.7	19.6' by 7.6'	Dra
UGC 8201	DDO 165	Im IV-V	13h06.3m +67°42'	14.1(B)	16.2(B)	3.5' by 2.5'	Dra

dusty, that has recently entered the M81 group's thin gas cloud, resulting in intense star formation. Spectra taken of M82's nucleus, in fact, closely resemble that of the Orion Nebula, another center of star formation. It has been proposed that this star-burst nucleus can be modelled as 100,000 Orion nebulae (Jones & Rodriguez-Espinosa, 1984).

Visually, M82 appears quite large: 7' by 1'.5 in p.a. 75°. The brightest region is about 1' in diameter and is just east of the center of the galaxy. High magnifications may be used on this well-defined object. At 165x, this high-surface-brightness object begins to show several prominent lanes cutting through its length at various angles. The nearly central bright spot is almost stellar with a tiny surrounding halo. The fall-off in brightness is very rapid around its periphery. With his 6-inch refractor at 140x, Brian Skiff saw a 2'.5 by 0'.5 core with two clearly visible dark breaks, with the easternmost of the resulting patches being the brightest.

Similar in appearance to the prominent Local Group galaxy M33 is the M81 group's **NGC 2403**. In *The Hubble Atlas of Galaxies*, author Allan Sandage (1961) chose to print photographs of the two side-by-side to point out the resemblance of these two Scd spirals. The content of this nearly face-on spiral galaxy is well documented. Catalogs exist for HII regions (Hodge & Kennicutt, 1985) and stellar associations (Hodge, 1983). Even though NGC 2403 is 3.25 Mpc distant, 19 globular cluster candidates have been detected in it (Battistini et al., 1984), although the brightest has a V magnitude of 17.9 and is out of range for most amateur instruments.

My first impressions of NGC 2403 came through my telescope's 9x60mm finderscope. The similarity with M33 was evident as the view reminded me of, say, a 5x24mm finder view of the Pinwheel. At 165x the 13-inch showed a very bright glow, 8' by 5', in p.a. 120°, with the brightest portion being a 2' diameter region around the center, with no sharp concentration to a nucleus. The spiral structure is discerned as a slightly brighter halo within the glow. I was unable to observe the origin of the ill-defined arms. The most striking feature of NGC 2403 is the presence of a dozen stars of 11th to 15th magnitude superposed over its face, many of these in front of the spiral arms. The brightest star is off the southeast edge of the galaxy. At 310x I tried for some or the stellar as-

NGC 4236 photographed by Martin C. Germano (8-inch f/5 reflector, hypered Tech Pan film, 75-minute exposure).

sociations from the previously mentioned article by Paul Hodge. Surprisingly, eight of these were visible with four associations even showing some size.

Probably few amateurs' lists of showpiece galaxies include **NGC 2366**. But this otherwise ordinary dwarf galaxy contains one of the most interesting HII regions known. The HII region, catalogued as **NGC 2363**, is in the same class of luminosity as such giants as the Tarantula Nebula and Eta Carinae. Most surprising is its presence in such a non-luminous dwarf galaxy. It appears that NGC 2363 is the center of an incredible burst of star formation (Kennicutt, Balick, & Heckman, 1980).

Through the telescope, the galaxy is as interesting as it is astrophysically. In the 13-inch it appears to measure 5' by 1'.5 in p.a. 45° with slight brightening at the southwest end around what initially appears at low powers to be a foreground star. At the northeast end, slightly out of the glow of the galaxy, lies a 13th-magnitude star. Initially ignorant of the HII region, I blinked the galaxy with an O III filter. The effect was stunning! The faint dwarf galaxy disappeared from view, but NGC 2363, the "star" on the southwest end, outshone everything else in the blackened field. A friend's wide-band nebula filter showed the same effect, but not so dramatically, since the emission of NGC 2363 is most intense in the O III lines. Increasing the magnification to 215x revealed a disk about 5" across that brightens substantially at the center.

NGC 2976 is located less than 2° southwest of M81 and is close enough to show effects of interaction with the latter. Radio observations have shown connecting bars of neutral hydrogen to M81's nearby satellite galaxies that are enhancements in the giant HI envelope permeating this entire region (Appleton, Davies, & Stephenson, 1981). Visually, NGC 2976 appears moderately bright, measuring 5' by 2' in p.a. 150°. It is uniform with the exception of a slightly fainter outer halo 1' wide. There is a 13th-magnitude star on the southwest edge. The southeast end of the oval glow seemed more rounded than the northwest end's sharper, more pointed appearance. High magnifications showed a nebulous patch 5" across due east of center at the edge of the haze. At this magnification slight brightening could be seen at the galaxy's center.

East of M81 in the same low-power field lies **NGC 3077**. This small galaxy also requires high magnifications. The 13-inch at 215x revealed a 3' by 2' haze oriented in p.a. 60°, brightening rapidly to the middle. Two 10th-magnitude stars were visible nearby, one 4' northwest and the other 10' west. A stellar nucleus with a bright round core 1' in diameter is centered in the oval glow. With his 6-inch refractor, Brian Skiff saw the core 45" across with a strong broad concentration. The brightest part of the galaxy appeared to be offset to the northwest side.

Observatory photographs of **IC 2574** show a wealth of detail in this irregular galaxy. Some of this detail is visible in small scopes. At 75x with the 13-inch, it appears about 9' by 4' in p.a. 30°, of fairly low surface brightness, and with plenty of foreground stars. A prominent bright region lies at the northeast end, which Brian Skiff observed with his 6-inch scope. This is one of the brightest HII regions in the galaxy, and in the 13-inch at 115x, it is a glow 1' in diameter with some central brightening. A nearly central 2' by 1' condensation in the galaxy's glow can be seen with some difficulty.

NGC 4236 closes out the list of easily observable galaxies in the M81 group. It is 15' by 3' in p.a. 165° and fairly faint. Some brightness variations could be seen across its surface, the brightest of which is a thin 5' by 1' bar displaced toward the north end of the galaxy.

The final seven galaxies featured required much patience

Further Reading

Appleton, P.N., R.D. Davies, and R.J. Stephenson. *MNRAS*, **195:**327, 1981.
Battistini, P., et al. *Astr. Astrophys.*, **130:**162, 1984.
Hodge, P. *PASP*, **97:**1065, 1983.
Hodge, P., and R. Kennikutt. *AJ*, **88:**296, 1985.
Holmberg, E. *Medd. Lunds. Obs.*, Ser. 2: No. 128, 1950.
Jones, B., and J.M. Rodriguez-Espinosa. *ApJ*, **285:**580, 1984.
Karachentseva, V.E., I.D. Karachentsev, and F. Borngen. *Astr. Astrophys. Suppl.*, **60:**213, 1985.
Kaufman, M. *S&T*, **73:**135, 1987.
Kennicutt, R., B. Balick, and T. Heckman. *PASP*, **92:**132, 1980.
Sandage, A. *The Hubble Atlas of Galaxies* (Carnegie Institution of Washington, Washington, 1961).
Solinger, A., P. Morrison, and T. Markert. *ApJ*, **211:**707, 1977.
Tully, R.B. *Nearby Galaxies Catalog* (Cambridge University Press, New York, 1987a).
Tully, R.B. *ApJ*, **321:**280, 1987b.
Tully, R.B., and J.R. Fisher. *Nearby Galaxies Atlas* (Cambridge University Press, New York, 1987)
Sargent, W.L.W., R. Sancisi, and K.Y. Lo. *ApJ*, **265:**711, 1983.

to be seen at all. They are interesting mainly because they are dwarf galaxies that can be detected at a great distance. Consulting the *POSS* before and after the observations confirmed what I was really seeing in some of these objects.

Holmberg II is the easiest of these remaining galaxies. It is round, about 5' across, and near an equilateral triangle of 13th- and 14th-magnitude stars. I was unable to pick out any of the HII knots visible on the *POSS*. **Holmberg I** is a very faint dwarf, visible in my telescope only at 115x with averted vision. It appeared less than 1' in diameter and round. On the *POSS* it is elongated 3:2 with its major axis oriented east-west. **M81 DW B** is another challenging galaxy that is less than 1' across on the *POSS*. In the 13-inch at 165x, the galaxy appeared almost 1' across and nearly uniform, except for a 20" diameter knot on the west edge. A magnification of 215x brings this feature out while the rest of the galaxy is not visible. Difficult but not impossible to see is **UGC 8201**. It was best at 115x, showing a 3' by 2' uniform glow elongated east-to-west.

Regarding observations of these final three galaxies, don't try seeing them at home! I could only see **UGC 4459** at 215x as a dull glow, 1' in diameter, with a 14th-magnitude star 1' east. **UGC 6456** appeared through my scope as a 1' diameter haze of extremely low surface brightness, best seen at 165x, while the *POSS* reveals some central brightening and a north-to-south elongation. After three frustrating attempts, I concluded that **M81 DW A** is not visible through my 13-inch telescope. It is in the same low power field as Holmberg II, although the two do not appear to be physically associated (Sargent, Sancisi, & Lo, 1983). In fact, it is practically an averted vision object on the *POSS*! It would be interesting to see if anybody has viewed this galaxy, from perhaps a high altitude site. After viewing these faint dwarfs, it is always fun to go back to something bright like M82 and nearly lose your dark adaptation!

The M81 galaxy group presents a wide range of objects, from the showpieces to the nearly invisible. But with some persistence, you can bring this latter class into visibility. As the group culminates at high altitudes for northern latitudes, it can be viewed through very little atmosphere on crisp winter nights. Test your observing skills on the M81 group on your next observing session.

A Portfolio of Galaxies for Backyard Telescopes

The Whirlpool Galaxy
photo by Jack Newton

M81 in Ursa Major
photo by Tony Hallas and Daphne Mount

M63 in Canes Venatici
photo by Bill Iburg

Dwarf galaxy Leo I
photo by Mace Hooley

M94 in Canes Venatici
photo by Jack Newton

The Andromeda Galaxy
photo by Tony Hallas and Daphne Mount

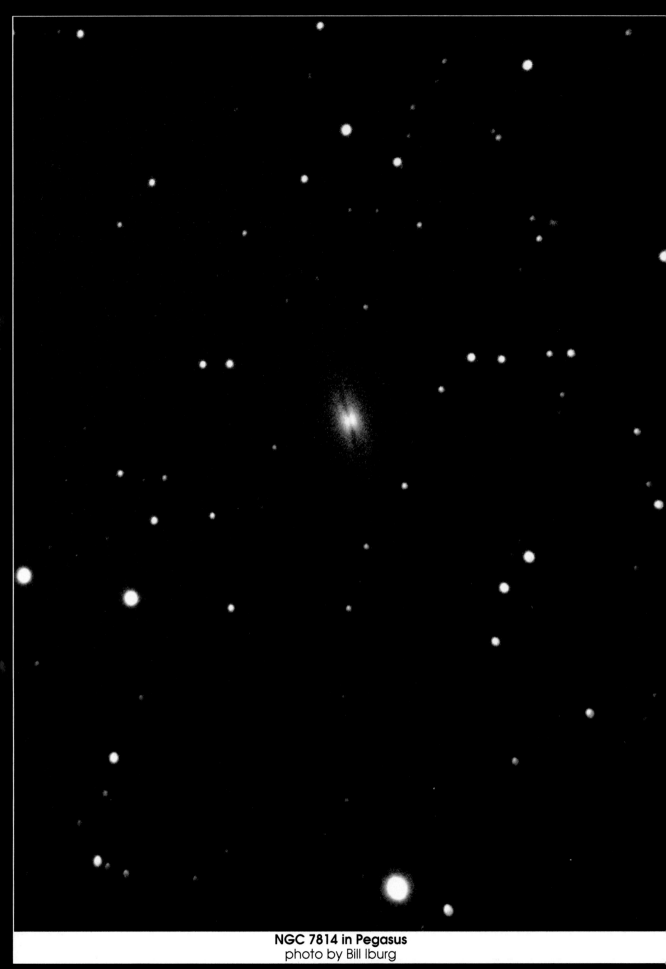

NGC 7814 in Pegasus
photo by Bill Iburg

M82 in Ursa Major
photo by Jack Newton

M105 in Leo
photo by Jack Newton

NGC 7479 in Pegasus
photo by Jack B. Marling

The Andromeda Galaxy
photo by Bill Iburg

Stephan's Quintet
photo by Jack Newton

NGC 891 in Andromeda
photo by Bill Iburg

NGC 404 in Andromeda
photo by Chris Schur

NGC 4631 in Coma Berenices
photo by Jack Newton

M106 in Canes Venatici
photo by Mace Hooley

NGC 6946 in Cepheus
photo by Bill Iburg

NGC 7331 in Pegasus
photo by Bill Iburg

The Andromeda Galaxy
photo by Frank Cathell

NGC 4485 (with supernova) and NGC 4490
photo by Jack Newton

NGC 2403 in Camelopardalis
photo by Jack Newton

ALL ABOUT M31

by Brian Skiff

It's in every astronomy book, along with Saturn, Mars, and the asteroids. Like the Moon, it is looked at and then overlooked. It is Messier 31, the majestic spiral galaxy in Andromeda, about which so much has been written, and from which we've learned so much about the behavior of the universe.

In the early pages of my observing log, I wrote: "It doesn't look like much in the telescope." I was observing from the middle of a large city, and was disappointed at not having seen spiral structure in my 2.4-inch refractor. M31 is not an easy object in which to view detail — many more clues to a galaxy's nature are visible in its face-on neighbor in Triangulum, M33, although M33 is half again as far from Earth. The reason why details in its spiral arms remain hidden is that M31 is highly inclined to our line of sight.

Yet sheer proximity has its advantages, like making the galaxy a bright naked-eye object. Staring at it in binoculars on a dark night, I've wondered what the view would be like were it not a narrow inclined oval, but a face-on, circular disk. Could we see the spiral arms in binoculars? I like to think so.

The Nature of M31

The most recent estimate of the distance to M31 is 760 kiloparsecs (nearly 2½ million light-years). The brightest stars in the galaxy are red and blue supergiants, distinguishable by their slight, irregular variability. If they are like stars of the same class in the Milky Way — such as Mu (μ) Cephei, a red supergiant, and Zeta^1 (ζ^1) Scorpii, a blue supergiant — and those in the nearby galaxies M33 and M101, they have absolute magnitudes of -8 to -9, and shine at apparent magnitude 15. With a large amateur scope, a few of the stars you see projected on M31 are actually inside the galaxy. Unfortunately no extensive study of these stars has been published, so finder charts are not yet available.

Without doubt other stellar members of M31 have appeared on amateur photographs — but only temporarily. They are novae, which appear at the rate of about two dozen per year. In 1956 the Mount Wilson astronomer Halton C. Arp published an early synopsis of novae in M31. Arp photographed the big spiral on 290 nights over a period of a year and a half, discovering 30 novae. The brightest of these shone at magnitude 15 or 16, corresponding to absolute magnitude -8.5 — close to that of bright novae in the Milky Way. Since the 1950s, the Russian astronomer A. Sharov and the Italian astronomer L. Rosino have followed novae in M31. Arp, Sharov, and Rosino all found the majority of novae within 15 arcminutes of M31's core, just as Milky Way novae are concentrated toward the galactic center in Scorpius and Sagittarius. Arp's paper, by the way, provides the only photoelectric measures of stars in the main body of M31, many of which are indicated on the photograph on page 10 and in the accompanying table.

As viewed from Earth, M31 is tilted only 12.5° from edge-on, making its spiral structure difficult to discern. But even at the turn of the century, before the galaxy's great distance was established, M31 was regarded as a "spiral nebula." Recently Paul Hodge of the University of Washington has found that the distribution of dust lanes and open clusters over the galaxy most strongly suggests a common two-arm spiral pattern wound about the nucleus several times. This outline is probably distorted to some extent by Messier 32, the compact elliptical companion galaxy apparently at the periphery of M31.

The dust lanes along the northwestern flank clearly show that this side is closer to us. The arms are spinning toward us on the northeastern end and away from us on the southwestern end. A 1970 radial velocity study by Vera Rubin and Kent Ford showed that HII regions extend out 2° from the galaxy's center, while radio 21-centimeter measures done by Morton Roberts and Robert Whitehurst, show that HI (neutral hydrogen) extends out to 2.5°. These data show that the rotation speed in the outer parts of the galaxy is constant, at around 230 km/sec — a common but not

---

Essential Facts

NGC 224 = Messier 31
  Right Ascension: 00h42m44.6s
  Declination: +41°16.1' (2000)
  Type: SA(s)b
  Dimensions: 177.8' x 63.1' (at B magnitude 25 per square arcsecond isophotes)
  V = 3.45   B-V = 0.91
  Mean Surface Brightness: 13.4 (magnitude per square arcminute)
  Inclination to line of sight: 77.5°

NGC 205
  Right Ascension: 00h40m21.7s
  Declination: +41°41.3' (2000)
  Type: E5 pec
  Dimensions: 17.4' x 9.8'
  V = 8.0   B-V = 0.84
  Mean Surface Brightness: 13.6

NGC 221 = Messier 32
  Right Ascension: 00h42m42.0s
  Declination: +40°51.9' (2000)
  Type: cE2
  Dimensions: 7.6' x 5.8'
  V = 8.2   B-V = 0.94
  Mean Surface Brightness: 12.3

sources: RC2, and de Vaucouleurs and Leach, 1981.

Shown here schematically (and on the following six pages) are the respective distance between and sizes of the M31 system (left) and the Milky Way system (on page 15). M31 itself spans 55.2 kiloparsecs — at its right tip is M32, lying at a distance of 5 kiloparsecs and measuring 600 parsecs across. At M31's left side is NGC 205, 7 kiloparsecs away and some 1.5 kiloparsecs across. Orbiting at a distance of some 90 kiloparsecs (below, overprinted on the type) are NGC 147 (first column) and NGC 185. The scale is 1 centimeter = 5.2 kiloparsecs. On this scale, M31's corona of dark material — its "missing mass" — would measure over a foot in diameter.

universal feature of rotation curves of other spiral galaxies including our own. For such systems to remain stable there must be great amounts of mass beyond where the luminous spiral arms end. At present we have nothing but this dynamical evidence to show that the matter exists — whatever it is, it just doesn't put out enough light for us to observe. The data imply, however, that as much as 90 percent of the masses of the Andromeda Galaxy and Milky Way are outside the traditional visible part of the objects.

The spiral arms in M31 are dominated by star clouds, or stellar associations — young, loosely bound collections of stars having a common origin. Nearly 200 of these regions, catalogued by the Canadian astronomer Sidney van den Bergh, are mapped in Hodge's *Atlas of the Andromeda Galaxy*. A number of the well-defined associations are marked on the photographic finder map on page 10; most have bright open clusters embedded in them.

The brightest and richest association is conspicuous enough to have its own NGC number. NGC 206, A78 on the map, is about 50′ southwest of the center of the galaxy. Van den Bergh made a photometric study of this star cloud and compared its characteristics to associations in our Galaxy. The brightest member stars are 17th magnitude objects, and more than 300 are brighter than about magnitude 21. Integrating the brightness of all stars above this limit, van den Bergh found a total V magnitude of 13.0 for the entire cloud. More recently, Hodge got a value more than a magnitude brighter. At the telescope, you might try to estimate the brightness of NGC 206 yourself. Use moderate or high powers to isolate it from the luminous background of the main galaxy.

NGC 206 does not seem to have any glowing H II regions within it — no "Orion Nebulae". But scattered all across the galaxy are many such objects. The French astronomer C. Courtes and his collaborators catalogued nearly one thousand emission nebulae in M31. Using an Hα filter to isolate the nebulae, their intriguing photographs give an unusual perspective of the galaxy. Curiously, none of the nebulae are as large or as bright as those in M33 and M101, which are easily viewed in a large telescope even though their parent galaxies lie much farther away. From my own observations and inspection of photographs, it seems that none of the M31 nebulae are conspicuous in amateur telescopes. Even the brightest one, located on the finder map around C205 and A69, is overpowered by the light of the star cluster.

Some 400 open clusters in M31 have been catalogued, but not well studied. These are only the very brightest in the galaxy; the fainter ones tend to be too small and faint to distinguish from stars. Hodge found that M31's clusters resemble the largest young open clusters in the Milky Way, such as the Double Cluster in Perseus and NGC 6231 in Scorpius. As with young clusters in our Galaxy, the luminous objects in M31 at least crudely outline the spiral arms.

While the open clusters are not well observed, the globular cluster system in M31 has been the subject of repeated and extensive study. The most recent is by David Crampton and his collaborators. Their catalogue of globulars includes 509 objects found on slitless spectra plates taken with Mauna Kea's 3.6-meter Canada-France-Hawaii telescope. The spectral data enabled them to confirm all the objects as star clusters: similar-looking H II regions, for instance, could be distinguished by their emission lines. They were able to derive colors and diameters for most of them as well.

The M31 system of globulars resembles the Milky Way's in many respects. The clusters concentrate in a halo around the center of the galaxy, and thin out away from the center at about the same rate as we see in our Galaxy. Likewise, clusters more distant from the centers of both galaxies have larger diameters, since gravitational forces are not so strong. The distant clusters are also intrinsically fainter, but the reason for this is not obvious.

The Andromeda Galaxy has eight known companion galaxies — all intrinsically faint ellipticals. The two brightest are familiar: Messier 32 and NGC 205 (infrequently labeled as Messier 110). M32 is a "compact elliptical," that is, its central surface brightness is very high despite being a small object, where a low central density would normally be expected. It is probably projected slightly in the foreground of M31, but is close enough to M31 — perhaps 5 kiloparsecs away — that both galaxies are distorted as a result. The photo of M32 in Arp's *Atlas of Peculiar Galaxies* shows a faint streamer extending from the main body of the galaxy.

NGC 205 has been well-studied. Its dark patches and anamalous association of blue stars near the center have been known about since Walter Baade's explorations of the system several decades ago. Detailed surface photometry shows that the outer regions of this galaxy are also distorted in the direction of M31. (It probably lies 7 kiloparsecs from M31.) A color-magnitude diagram of its individual stars was recently published which

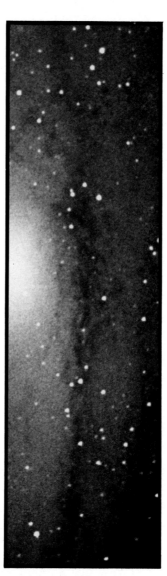

Under dark skies, 4-inch or larger telescopes reveal M31's most prominent dark lane, lying on the northern side of the nucleus. Photo by Martin C. Germano; 8-inch f/10 SCT, 60-minute exposure on 103a-F.

Below: A photographic finder chart for stellar associations, globular clusters, open clusters, and stars in M31 and in its field. North is up, east at right in the photo — V magnitudes, colors, dimensions, and/or notes are given for various objects in the accompanying table. Photo by Brian Skiff, who used the 13-inch astrographic "Pluto" camera at Lowell Observatory's Anderson Mesa Station. The exposure lasted 30 minutes on 103a-O emulsion. *Right*: With a 10-inch f/5 Newtonian, Lee C. Coombs made two identical 15-minute exposures on 103a-O to make this composite of M31.

Associations

A 29 — contains a prominent HII region
A 40 — compact
A 41 — compact
A 42 —
A 49 —
A 54 — rich star cloud
A 67 — contains two emission knots corresponding to C223-224
A 69 — contains bright blue stars and a triangular emission region with a very bright blue central star (C205)
A 78 — NGC 206, richest star cloud in the galaxy. The brightest stars are blue main sequence stars at V~17.4. The integrated magnitude is B=11.9, derived from the sum of isophotal contours.
A102 — large; many subclusters

Globular Clusters

Name	V	B-V	diam. (FWQM)	conc.
G 35	15.6	0.89	2.9"	m
G 52	15.7	1.00	2.3	c
G 64	15.1	0.78	2.3	c
G 72	15.0	0.96	2.2	m
G 73*	15.0	0.83	—	m
G 76	14.3	0.84	3.6	m
G 78	14.3	1.13	3.2	m
G 87	15.6	0.84	2.9	m
G 96	15.5	0.90	2.7	m
G119	15.0	0.82	2.7	c
G156	15.6	0.90	2.5	m
G172	15.2	0.89	2.4	c
G205	14.8	1.27	2.9	m
G213	14.7	0.90	2.5	c
G222	15.2	0.98	3.2	m
G226	15.5	0.98	3.8	m
G229	15.0	0.79	3.4	m
G230	15.2	0.84	2.9	m
G233	15.4	0.96	2.6	m
G244	15.4	1.00	2.6	m
G254	15.7	0.96	—	m
G257	15.1	0.87	3.2	m
G272	14.8	0.86	3.4	m
G279	15.4	0.74	4.9	d
G280	14.2	0.95	2.7	m
G286	15.7	0.78	2.5	m
G287	15.8	0.84	2.2	c
G302	15.2	0.78	2.5	m
G305	15.6	0.98	2.2	m
G315	15.7	0.85	—	m

*G73 is associated with NGC 205.

Open Clusters

C179 — involved in A22
C202 — involved in A24
C203 — involved in A24
C205 — involved in A69
C223 — involved in A67
C224 — involved in A67
C259 — involved in A29
C270 — involved in A32
C275 — involved in A59
C311 —
C312 — involved in A48
C313 —
C410 — involved in A33

Stars

Name	V	B-V
A	13.0	0.64
B	12.1	0.92
C	12.0	0.56
D	12.2	0.26
E	12.7	0.72
2a	15.4	0.92
2b	15.4	0.80
23a	14.4	0.43
23b	15.3	0.75
29a	14.7	0.55
30a	14.2	0.55
30b	14.8	1.28
30c	16.5	0.99

showed a distance indistinguishable from that of M31, and the characteristics of the stars and that its stars are similar to stars in Milky Way globulars. The resemblance to old globular clusters indicates that most star formation in NGC 205 occurred in the ancient past. But several dark clouds and a patch of young blue stars near the center show that star formation is probably going on right now, though at a very low rate. In this respect NGC 205 resembles NGC 185, one of two other companion galaxies about 7° north of M31. It and NGC 147 are somewhat fainter objects, having fewer stars and less mass.

Four more dwarf elliptical galaxies were found by van den Bergh during a special photographic survey using Schmidt plates. These objects, designated Andromeda I, II, III, IV, were resolved on photographs taken with the Palomar 5-meter telescope. They are approximately at the distance of M31, but because of their exceedingly low surface brightness, they are invisible in amateur instruments.

Observing Andromeda's Great Spiral

Messier 31 is an easy naked eye object, visible even on hazy nights, in full moonlight, or from slightly light-polluted skies. In relatively poor conditions, though, you can make out only a roughly circular spot. Under dark skies, its elongation is striking in contrast to the starry foreground. From outside Flagstaff, I can see a faint star — about magnitude 6.9 — near the southwestern tip. Under the best conditions the galaxy extends past this star, which implies an observed length of about 3½°.

A 2.4-inch refractor at low power (25x, say) shows the halo more than 2° long in position angle 40°, but the outer regions fade very

Above: Jamey Jenkins used a 4¼-inch reflector and 103a-O film to make this sequence of four frames of progressively less exposure time. The first is a 5-minute exposure, showing an overexposed nuclear region and two dust lanes. The second and third frames — 2-minute and 1-minute exposures — still offer an overexposed nucleus. But the fourth frame, for which the camera opened just a few seconds, shows M31's starlike nucleus. *Far right*: A closeup of M31's center, showing perhaps the few inner kiloparsecs of the core, by Richard Berry.

slowly and really have no definite edge. The galaxy contains some weak, broad condensation, but the core appears smooth and well concentrated. As you scan toward brighter and brighter areas in the center, the core grows progressively more eccentric in relation to the halo — the northwest side of the core fades more abruptly than the southeast side, placing the nucleus off-center to the northwest. Some stars appear superimposed on the galaxy, including a 12th magnitude one on the south-southwestern side of the core.

Further details become visible in larger aperture scopes. With a 6-inch, the halo measures about 120' x 20' with a pronounced core 10' across. The dark lane northwest of center can be seen without difficulty, showing some faint haze beyond it. In a 10-incher the halo extends to about 40' wide. The core measures 7' in width, cut off by dark lanes on the minor axis, but its length is indefinite since it fades evenly along the major axis. The inner 2.5' of the core is quite circular and appears "opaque" like other unresolved objects of high surface brightness, such as the centers of Messier 87 and Messier 49 in Virgo. The brightness rises evenly to a central tip about 10" across.

The prominent dark lane on the northwest side is a sharply defined strip about 1' wide as it passses only 5' from the core, but is clearly visible as it extends southwest past two stars aligned northeast-southwest about 15' from the center. Two more indistinct dark lanes are discernable, one further northwest and one along the southeastern flank. On the western side of two bright stars lying about 15' northeast of the center a spike of dark material intrudes toward the nucleus. In larger apertures, the bright

The Andromeda Satellites

M32
is easily visible in the same low-power eyepiece field as M31. It appears as a small round patch of bright light, and doesn't change its appearance much no matter what telescope you use on it. Photo by Martin C. Germano.

NGC 205
is the other satellite galaxy nearby M31. It appears as a faint, oval smudge of light without the strong central condensation of M32. It too lies in the same low-power field, though farther north. Photo by Lee C. Coombs.

NGC 147
is way up in Cassiopeia, some 7° north of M31 itself. It is one of the least massive galaxies known, and not surprisingly is difficult to spot. But a 12-inch scope should show it as an even patch of light. Palomar Observatory photograph.

NGC 185
also lies in Cassiopeia, and is slightly smaller and denser than its neighbor NGC 147. On dark nights it is just visible in a 12-inch scope — but it is certainly not easy. Palomar Observatory photograph.

nucleus has a nearly stellar center, but there also appear to be some odd brightenings with radii of about 15".

In October 1982, I surveyed M31 with a 10-inch Ritchey-Chretien telescope, looking particularly for globular clusters and stellar associations. At 200x the limiting V magnitude was 15.3 — I determined this afterward when I found that a "threshold" star mentioned in my notes has been measured photoelectrically by Arp (this is star 23b in the Table). Nearly all the accessible objects lay along the outer rim of the galaxy, and not in the inner halo or core.

A good place to start detailed observing of M31 is with the bright star cloud NGC 206 (A78 in the finder map), embedded in the southwestern arms. In the 10-inch Ritchey at 150x it was conspicuous as a 2' x 1' haze elongated north-south with a sharply defined eastern flank. A few faint mottlings, probably mostly foreground Milky Way stars, scatter over the surface. Along the major axis of the region the light merges smoothly into the general haze of the galaxy's spiral arms. The globular G76, one of the brightest in the galaxy, lies not far to the southeast, close by an oblique triangle of stars. The cluster is just separable from a star of similar magnitude immediately southeast of it.

Heading northeast toward M32, the open clusters C202 and C203 — separated by 16" — are visible and seem nonstellar at high power when compared to faint stars about 2' southeast. Midway along the southeastern rim is C410, just visible as a starlike spot. G229, set against the outer reaches of the core, is difficult to distinguish, but is just visible at 150x. Moving northeast, A40 is barely visible as a faint haziness associated with a small triangle of unequally bright

The Andromeda Galaxy photographed with a backyard Schmidt camera, showing the full extent of the visible disk and the two nearest satellite galaxies, M32 and NGC 205. The knot in M31's spiral arm, at upper right, is the giant starcloud NGC 206. Photo by Ron Potter; 8-inch f/1.5 Schmidt, 4-minute exposure on hypersensitized TP 2415.

stars (I saw the southwestern member only with averted vision). Just visible 5' north-northeast is A42.

An intriguing feature marked on the *Atlas of the Andromeda Galaxy* is the string of open clusters C311-C312-C313. Disappointingly, they were not visible in my telescope. When the bright stars northeast and nearby southeast of the clusters were centered in a 30' field, I noticed that the brightness gradient from the galaxy's disk to the sky was very strong, perhaps more conspicuous here than anywhere else.

Far out in the northeastern extremity, A102 is discernable as a faint stellar spot of about 14th magnitude, with some faint haze around it. This feature is visible probably only because the background has faded out completely.

Turning southwest along the obscured flank, few distinct objects are visible. Out of the way of most of the dark matter is A54, a broad mottled area I viewed best at medium power (about 100x) in the 10-inch scope. But the small association A67 was visible only at 200x, as was the cluster G78.

Messier 32 is the most obvious companion to the Andromeda Galaxy. A 2.4-inch telescope easily shows it at the edge of M31's halo, 24' due south of the nucleus. It appears tiny next to its parent. The edges are irregular, and it brightens suddenly to a sharp stellar nucleus. In a 6-incher it is about 3'x2', elongated southeast-northwest with a stellar nucleus. The intense core has the character of a planetary nebula. Viewed in a 10-inch instrument, the halo elongates toward the nucleus of M31. The brightness rises smoothly in the outer parts, then suddenly to a nearly stellar nucleus, but there is no distinct core. Using a 12-inch Cassegrain

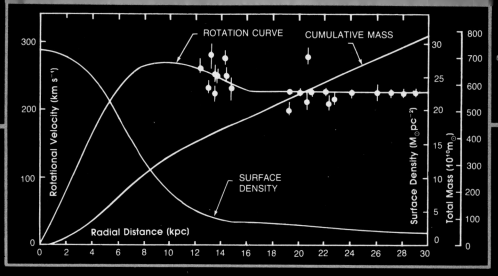

This graph shows the relationship between M31's cumulative mass, its surface density, and its rotation curve. Moving outward from the nucleus, surface density drops quickly, while cumulative mass slowly rises. The rotation curve peaks at a radius of about 8 kiloparsecs, and levels off: this indicates huge amounts of dark matter lying far beyond the visible disk of the galaxy. After E.S. Light, R.E. Danielson, and M. Schwarzschild, Princeton University Observatory.

M31 Through the Eyepiece

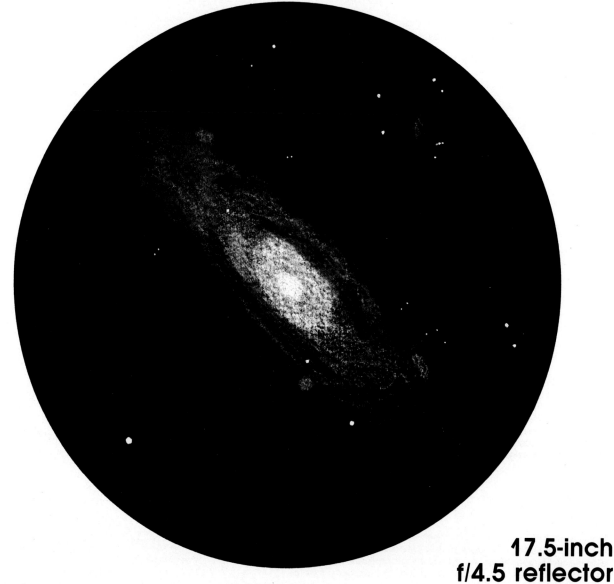

17.5-inch f/4.5 reflector at various powers

Composite sketch by Dave Eicher

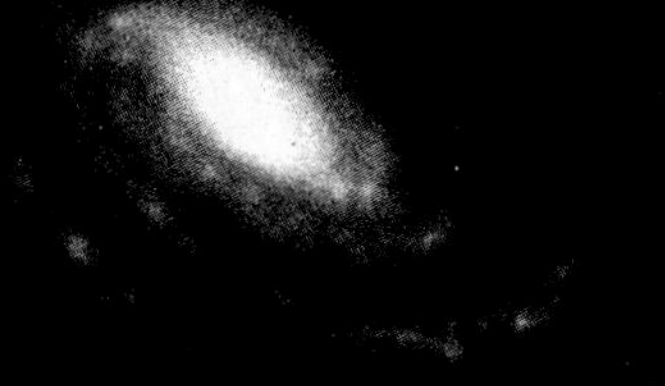

The distance to M31 is 760 kiloparsecs, or nearly 2½ million light-years. On these pages the distance between M31 (page 9) and the Milky Way (right) is 1.3 meters, and the sizes and orbital distances of the galaxies are drawn to scale. Surrounding the Milky Way are the two Magellanic Clouds (below, overprinting on text), which orbit at distance of 53 (Large) and 60 (Small) kiloparsecs. The Milky Way measures 35 kiloparsecs across. The scale for this model of the M31/Milky Way pair is 1:163,990,000,000,000,000,000,000,000!

telescope, my friend Chris Luginbuhl finds the core elongated in pa 160°, about 1.75′ x 1′ in extent. The faint halo merges with that of M31, so the total size is difficult to estimate. The nucleus is less than 10″ across, but clearly nonstellar. A faint star is visible 2′ east of the center.

This second companion lies 37′ northeast of the center of M31. Its comparatively large angular size results in a moderately low surface brightness. Still it is visible in binoculars; small instruments show the galaxy as a diffuse haze elongated southeast-northwest and weakly conentrated toward the center. A 2.4-inch refractor will catch two stars about 4′ and 6′ south of the center. In larger apertures it is best viewed at medium powers. With my 10-inch Ritchey it was about 9′x3′ in size, elongated in p.a. 165°. The core is 3′x2′, but no sharp nucleus is visible. A 14th magnitude star is involved in the halo toward the south-southeastern end.

The finder chart shows all the objects discussed above plus many others that should be visible in fairly large amateur instruments, based on my inspection of many photographs.

Projects in Observing M31

There are a few areas in which readers can make some really neat observations of M31. These either haven't been done at all or haven't been discussed in popular astronomical magazines for many years.

The first is redetermining the maximum visual extent of the galaxy. We've all read about being able to see it 5° long. Has anyone *tried* this lately? Such observations should be referred to specific stars in the field that can be measured later on photographs. Perhaps it would also be interesting to use different instruments from the same site to see how the observable length changes with different apertures and magnifications.

Experimenting with nebular filters, visually and photographically, may turn up some interesting results. I encourage someone to try photographing the galaxy with a narrow-band filter — an Hα or a Lumicon UHC — to capture emission nebulae. On such photographs the general light from the galaxy would be greatly suppressed, but that should make the nebulae all the more obvious. As noted in Table II, some of the associations are covered by HII regions, so these would be prime targets for such experiments. Owners of large telescopes might see if some of these nebulae pop out visually. I have tried it successfully on M33 and M101, but have not yet had the opportunity on M31.

In going through the technical literature, I noticed that reports on novae by Rosino and Sharov suddenly ended in the mid-1970s. If they are no longer monitoring the galaxy for novae, why can't amateurs take over? A big percentage of the ones that occur (one or two every month!) are brighter than 17th magnitude, within easy photographic reach of a 10-inch or 12-inch telescope. Such searches could be done by using "Problicom"-type projection blinking techniques or with a direct blink comparator. Will the next nova in M31 be discovered by backyard astronomers?

Brian Skiff is an astronomer at Lowell Observatory in Flagstaff, Arizona, where he is working on photometry of solar-type stars and discovering asteroids. He is an experienced deep-sky observer and a long time contributor to Deep Sky. *He is currently working on a book on observing, to be published by Cambridge University Press in 1985.*

An Observer's Bibliography for The Andromeda Galaxy

Arp, Halton C. 1956. Novae in the Andromeda Galaxy. *A.J.* **61**:15

Van den Bergh, Sidney. 1964. Stellar Associations in the Andromeda Nebula. *Ap. J. Suppl.* **9**:65.

van den Berg, Sidney. 1966. The Association OB78 in the Andromeda Nebula. *A.J.* **71**:219.

Courtes, C., Maucherat, A.J., Monnet, G., Pellet, A., Simien, F., Astier, N., and Viale, A. 1978. A Survey of H II Regions in M31. *Astr. & Ap. Suppl.* **31**:439.

Crampton, D., Cowley, A., Schade, D., and Chayer, P. 1984. The M31 Globular Cluster System. *Ap. J.* (in press).

Hodge, Paul W. 1979. The Open Star clusters of M31 and Its Spiral Structure. *A.J.* **84**:744.

Hodge, Paul W. 1981. *Atlas of the Andromeda Galaxy.* University of Washington Press: Seattle.

Hodge, Paul W., and Kennicutt, Robert. 1982. High-Resolution Optical Surface Photometry of M31. *A.J.* **87**:264.

Light, E.S., Danielson, R.E., Schwarzschild, M. 1974. The Nucleus of M31. *Ap. J.* **194**:257.

Mould, Jeremy, Kristian, Jerome, and Da Costa, Gary. 1984. Stellar Populations in Local Group Dwarf Elliptical Galaxies. II. NGC 205. *Ap. J.* **278**:575.

Roberts, Morton, and Whitehurst, Robert. 1975. The Rotation Curve and Geometry of M31 at Large Galactocentric Distances. *Ap. J.* **201**:327.

Rubin, Vera C., and Ford, W. Kent, Jr. 1970. Rotation of the Andromeda Nebula froma Spectroscopic Survey of Emission Regions. *Ap. J.* **159**:379.

Sharov, A.S., and Lyuty, V.M. 1983. Photoelectric Catalogue of Globular Clusters in the Andromeda Nebula and its Companions NGC 147, NGC 185, and NGC 205. *Ap. & Space Sci.* **90**:371.

de Vaucouleurs, G., and Leach, R. 1981. Revised Coordinates of the Nuclei of M31 and M33. *P.A.S.P.* **93**:190.

de Vaucouleurs, Gerard, de Vaucouleurs, Antoinette, and Corwin, Harold G. 1976. *Second Reference Catalouge of Bright Galaxies* (RC2). University of Texas Press.

Vetesnik, M. 1962. Photographic Photometry of Star Clusters in the Galaxy M31. *Bull. Astro. Inst. of Czech.* **13**:180.

Exploring the Region of M51
by Robert Bunge

On June 8, 1907, French observer Guillaume Bigourdan recorded his last observation of M51 for his visual survey of Dreyer's *New General Catalogue (NGC)*, which Bigourdan titled *Nebuleuses et D'Amas Stellaires*.

Bigourdan had set out to observe all of the NGC objects using a 12-inch refractor at the Paris Observatory. Bigourdan recorded M51, the famous Whirlpool Galaxy in Canes Venatici, as a brilliant, 50"-diameter nebula with an even more brilliant central region. Apparently Bigourdan did not see the galaxy's spiral arms. His summary was drawn from observing M51 four times during the course of the survey. Unlike most observers of his day, Bigourdan looked at more than just M51 and its companion, NGC 5195. He observed at least five other NGC galaxies in the region around M51. Bigourdan looked at the NGC objects and didn't do a detailed search for any other nearby galaxies. If he had done so he might have noticed several other galaxies in the region.

Over the past century things haven't changed much. Today amateurs by the thousands enjoy viewing M51 but rarely, do they bother to find the dozen or so fainter galaxies within a two or three degree circle around the Whirlpool galaxy. Many of these galaxies are visible with modest backyard telescopes.

Before you attempt to observe these galaxies, however, do take a look at M51. In the 31-inch F/7 Newtonian telescope at Warren Rupp Observatory the spiral arms of M51 can be seen with ease by visitors who have never looked through a telescope before. To the trained eye the Whirlpool breaks down into a wealth of detail that is almost impossible to draw on paper. Knots and twists in the arms burst through the glow of the galaxy's disk of stars that spread out to form what appears to be the faint bridge that connects the large galaxy NGC 5194 with its small companion NGC 5195.

My interest in hunting the galaxies near M51 was inspired by seeing the faint edge-on galaxy that appears in many long-exposure photographs of M51 (on the cover of volume one of *Burnham's Celestial Handbook*, for example) just to the northeast side of NGC 5194. Despite trying repeatedly to see it repeatedly with the 31-inch scope, it remained invisible.

I discovered the galaxy's designation is **IC 4277** and found that it receives only a listing in *NGC 2000.0* — no magnitude or description. IC 4277 and nearby **IC 4278** were both discovered by American astronomer James Edward Keeler on photographs of M51 that he took using the 36-inch Crossley reflector at Lick Observatory.

After an unsuccessful try at IC 4277 I decided to try to find some of the NGC objects plotted on chart 76 of *Uranometria 2000.0*. Moving from IC 4277 almost directly south, just less than a degree, I found **NGC 5198**. At 200x this galaxy showed little detail other than its shape and a stellar core. Its magnitude is approximately 12 and the galaxy's longest dimension is 2.1'. John Herschel was the source of Dreyer's visual

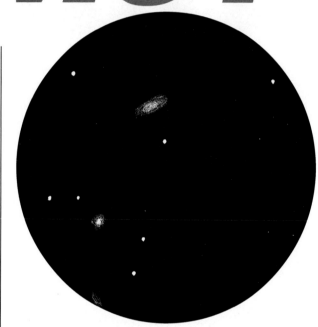
NGC 5169 (top), NGC 5173 (middle) and MCG +08-25-006 sketched with a 31-inch f/7 reflector at 200x. All sketches in this article are by Robert Bunge.

NGC 5198 sketched with the 31-inch f/7 scope at 200x. Opposite page: M51 photographed by Ron Potter with a 14-inch f/11 SCT, hypered Tech Pan, and a 120-minute exposure.

62

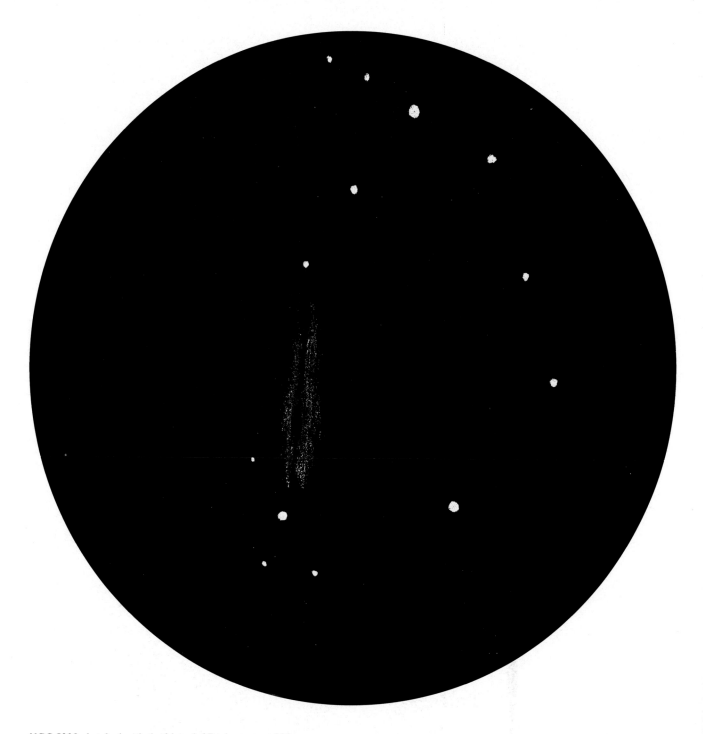

NGC 5229 sketched with the 31-inch f/7 telescope at 200x.

description, which called this object "pretty faint, pretty small, round and much brighter in the middle." When he observed it in 1897 and 1907 with the 12-inch refractor, Bigourdan thought NGC 5198 was round, notably brighter toward the center, and about 40" in diameter.

About two arcminutes west are **NGC 5169** and **NGC 5173**. At 200x both of these galaxies lie within the same field. NGC 5169, the northernmost of this pair, was a fine sight with averted vision. This galaxy appeared as a large, faint oval with a brighter central region that did not show any sort of stellar core. Some distance to the south-southeast lies an anonymous galaxy that is not plotted on *Uranometria 2000.0* or listed in the NGC or IC catalogs.

The *Revised New General Catalogue (RNGC)* lists NGC 5169 as having a photographic magnitude of 14.5, while the more precise *Uppsala General Catalogue (UGC)* offers a photographic magnitude of 13.6. *NGC 2000.0* splits the difference at photographic magnitude 14. In any case, small backyard telescopes under good conditions show this galaxy. Bigourdan saw it with the 12-inch refractor and described it as being about 25" or 35" across with a bright center surrounded by a trace of faint nebulosity. Dreyer described NGC 5173 as very faint, pretty small, and round. Bigourdan thought it was "round, with a very feeble surrounding nebula and about 20" to 30" in diameter." The 31-inch at 200x easily showed Bigourdan's surrounding glow and a bright nucleus with averted vision.

The anonymous galaxy at the southern edge of the field of view later proved to be **MCG +08-25-006**. Of all the major

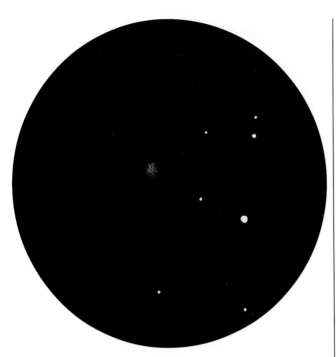

UGC 8588 sketched with the 31-inch f/7 scope at 200x.

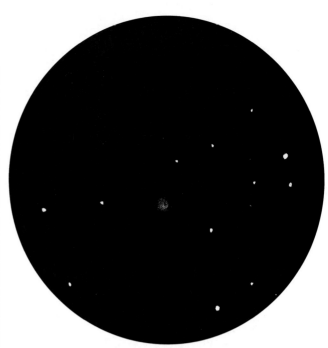

UGC 8538 sketched with the 31-inch f/7 reflector at 200x.

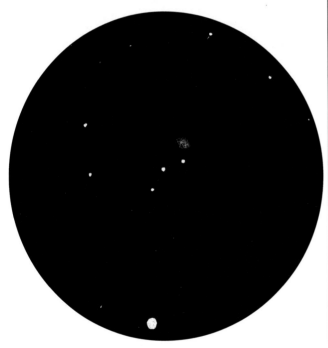

MCG +08-25-010 sketched with the 31-inch f/7 scope at 200x.

MCG +08-25-017 sketched with the 31-inch f/7 scope at 200x.

sources, only Vorontsov-Velaminov's *Morphological Catalogue of Galaxies (MCG)* listed this galaxy. It was easy to see in the 31-inch scope as a faint, north-south elongated nebulosity. I could not see any other detail.

To finish off most of the nearby NGC galaxies I moved the telescope slightly northeast of M51 to **NGC 5229**. After centering the location of the galaxy in the finder scope, I looked in the eyepiece of the 31-inch scope and was greeted by the sight of a large, beautiful edge-on galaxy with a broad equatorial dark lane. NGC 5229 is one of about 900 objects discovered by American astronomer Lewis Swift with the 16-inch Clark refractor at Warner Observatory in New York. Known for his many comet discoveries, Swift also searched for nebulae. Based on Swift's description, Dreyer listed NGC 5229 as "extremely faint, large, much extended and very difficult." The galaxy shows no hint of a nucleus or other brightening toward its center in the 31-inch. In 1901 Bigourdan described NGC 5229 as a large, diffuse nebula measuring 1' in diameter. The UGC lists NGC 5229 as an edge-on SBc system with a photographic magnitude of 14.5. The RNGC description reads: "Edge on, little brighter towards the middle, very flat, with an equatorial dark lane."

In the course of looking at these galaxies I had seen several other objects not plotted on *Uranometria 2000.0* in the area around NGC 5198 and the NGC 5169/NGC 5173 pair. The next step was to examine in detail a copy of the POSS chart of the region to see what else I might be able to observe in the area around M51.

Just south and slightly east of NGC 5198 I found **MCG +08-**

List of Galaxies near M-51

Object Name	Ra (2000)	Dec	Mag.	P
UGC 08320 (ZWG 245.036) (MCG+08-24-093)	13 14.4	+45 55	14.0 (Z)	N/O
UGC 08331 (ZWG 245.038) (MCG+08-24-097)	13 15.5	+47 29	15.6 (Z)	N/O
IC 4257	13 27.3	+46 52	N/A	
NGC 5169 (UGC 08485) (ZWG 246.002) (MCG+08-25-004)	13 28.2	+46 40	13.6 (Z)	
NGC 5173 (UGC 08468) (ZWG 246.003) (MCG+08-25-005)	13 28.4	+46 36	13.5 (Z)	
MCG+08-25-006	13 28.6	+46 30	16.0 (M)	
IC 4263 (UGC 08470) (ZWG 246.004) (MCG+08-25-007)	13 28.6	+46 56	15.4 (Z)	
MCG+08-25-009 (ZWG 246.005)	13 28.8	+46 15	15.1 (Z)	
MCG+08-25-010 (ZWG 246.006)	13 29.2	+46 31	15.7 (Z)	
UGC 08499 (ZWG 246.007) (MCG+08-25-011)	13 29.7	+45 24	15.2 (Z)	N/O
NGC 5198 (UGC 08499) (ZWG 246.010) (MCG+08-25-015)	13 30.2	+46 40	13.2 (Z)	
IC 4277	13 30.3	+47 19	N/A	UN
IC 4278	13 30.4	+47 15	N/A	UN
IC 4282	13 31.3	+47 11	N/A	N/O
MCG+08-25-017 (ZWG 246.011)	13 31.6	+46 10	15.0 (Z)	
UGC 08538	13 33.2	+45 52	17.0 (U)	
NGC 5229 (UGC 08550) (ZWG 246.013)	13 34.1	+47 55	14.6 (Z)	
UGC 08588 (ZWG 246.016) (MCG+08-25-023)	13 35.8	+45 56	15.1 (Z)	
UGC 08597 (ZWG 246.018) (MCG+08-25-025)	13 36.3	+46 13	15.2 (Z)	N/O
UGC 08601 (MCG+08-25-017)	13 36.5	+47 45	17.0 (U)	N/O
UGC 08611 (ZWG 246.020) (MCG+08-25-029)	13 36.9	+44 45	15.5 (Z)	N/O
MCG+08-25-030	13 37.7	+48 14	16.0 (M)	N/O
NGC 5256 (UGC 08632) (ZWG 246.021)	13 38.3	+48 17	14.1 (Z)	

I have tried to cross reference names of objects whenever possible. Positions attained from sources in the following order: NGC2000, Zwicky, UGC, MCG. Photographic magnitudes attained from sources in the following order: Zwicky, UGC, MCG. N/O in last column means Bunge did not try to observe object. UN means unsuccessful observation.

Chart by Robert Bunge

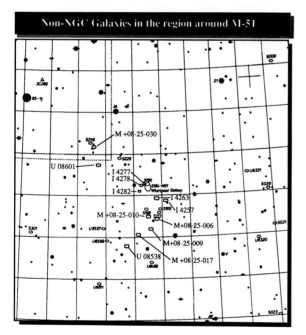

MCG +08-25-009 sketched with the 31-inch f/7 scope at 200x.

25-010, also the fifth galaxy listed in region 246 of Fritz Zwicky's *Catalogue of Galaxies and Clusters of Galaxies (CGCG)*. The CGCG lists the galaxy as having a photographic magnitude of 15.7, while the MCG provides dimensions of 6" by 5" and an inner core with a diameter of 1.5" across. I observed the galaxy in the 31-inch at 200x as a round, featureless, even glow that was very difficult to see and was visible only with averted vision.

South and east of M51 lie two more IC objects that were photographically discovered by Keeler with the Crossley reflector. The easternmost is **IC 4263**. I found it to be a large, cigar-shaped object with no detail that was difficult to see even with averted vision. Dreyer affixed the following description to IC 4263: "extremely faint, pretty large, and much extended with a much brighter middle region."

The *UGC* gives it a photographic magnitude of 15.4 with blue-light dimensions of 1.9' by 0.4' (p.a. 105°).

As a study in contrast, neighbor **IC 4257** was a very small, almost stellar point with only a little fuzziness in the 31-inch scope. Too small to be listed in the *UGC*, IC 4257 fails to make any of the modern catalogs. *NGC 2000.0* lists the galaxy but does give a magnitude. From the Keeler photographs Dreyer described it as extremely faint, small, round, and diffuse. Since I found it to be small but bright, I suggest you use high magnification and study each star in the plotted region with great care for faint nebulosity surrounding it.

South of NGC 5169 and NGC 5173 you may be able to find **MCG +08-25-009**. This galaxy was easy to see in the 31-inch scope. With a photographic magnitude of 15.1 this

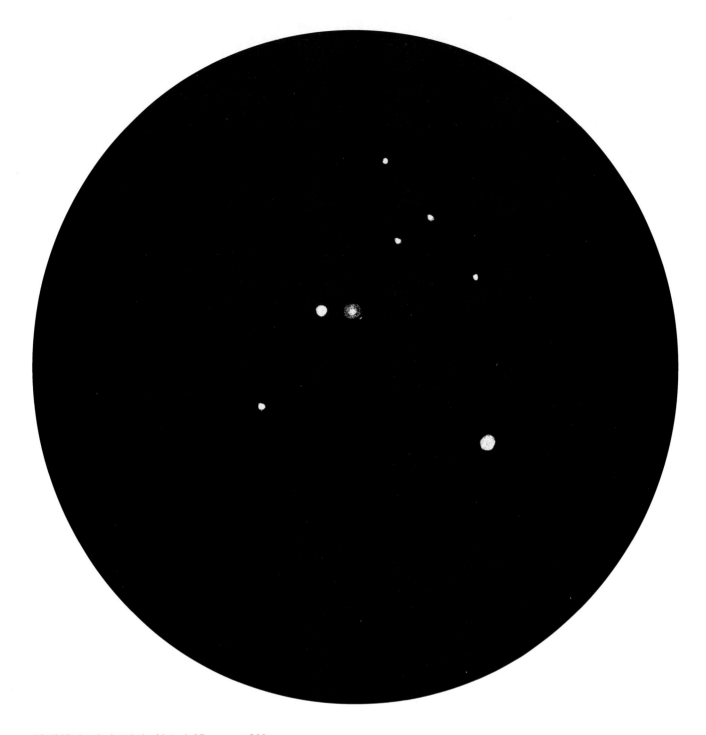

IC 4257 sketched with the 31-inch f/7 scope at 200x.

object has an oval shape with a bright, direct-vision stellar core. Averted vision really made this galaxy look good, appearing like a bright nucleus with a faint outer shell. Continuing southeast we find **MCG +08-25-017**, an object with a photographic magnitude of 15. This galaxy was also easy in the 31-inch scope at 200x. I saw it with direct vision as a large, round patch with a slight brightening toward the center. The MCG lists it as having a 3' by 2.5' central region and a size of 6' by 7'.

Yet another bump south and east brings you to **UGC 8538**. I looked for this galaxy for some time before I finally noticed it with 340x. I confirmed the galaxy's presence by moving the telescope slightly and by taking a short break and coming back to it only to still see it. I wasn't surprised to see the object listed at photographic magnitude 17 in the *UGC*. One must remember that *UGC* magnitudes fainter than 15.7 were estimated from the *POSS* photos, so it may in actuality be brighter.

The last galaxy I observed was **UGC 8588**, which is plotted in *Uranometria 2000*. This galaxy has a photographic magnitude of 15.3 and was a real challenge to see in the 31-inch scope at 340x. I saw no detail, only a very small, round glow. The *UGC* lists this galaxy as a dwarf spiral system with a blue-plate diameter of 1.4' and an inclination of 1.0, meaning that it is face-on to our line of sight.

Next time you're under dark skies give these galaxies a try. The challenge of finding some faint objects will never be greater than with this group of small, faint galaxies that surround the old Whirlpool.

Robert Bunge is an a reporter at the Springfield News-Sun *in central Ohio.*

The Galaxies of Canes Venatici

by Max Radloff

No matter what size your telescope is, many galaxies your telescope will show you will be dim little ovals with brighter middles. If you are getting a little bored with these faint fuzzies and would like to view some galaxies that are both large and bright, the sometimes overlooked constellation of Canes Venatici offers a variety of prominent galaxies with interesting structure, including a number of interacting pairs. Even if you don't own a large-aperture telescope, there is much you can enjoy here, because you can see many details with scopes of 8-10 inches in aperture if you sharpen your observing skills and push your eye and your instrument to the limit.

Layers of Galaxies

Many of the brightest galaxies in the sky belong to the long string of galaxy groups that stretches from the Sculptor group through our Local Group out to the heart of the local supercluster in Virgo. All of the brighter galaxies in western Canes Venatici are in this string, and they are in three distinct layers projected onto the same area of the sky. Closest to us at a distance of 5 megaparsecs (assuming $H_0=75$) is the CVn I cloud that contains M94 and M106 and NGCs 4214, 4244, 4395, and 4449. Behind this at almost twice the distance are both the CVn II cloud and the M101 group. The former contains NGCs 4111, 4242, 4290, 4618, 4631, and 4800, and the latter has M51 and M93 as outlying members. In the third level at 55 megaparsecs are the various parts of the UMa I cloud including NGCs 4145, 4151, 4217, 4369, and perhaps 5005 and 5033. The rich group in the eastern part of the constellation is much more distant at 30 megaparsecs, roughly twice the distance of the Virgo cluster.

Observing Galaxies

I suspect that my experience as a galaxy observer is like that of most amateurs: I was initially disappointed at seeing a series of faint and featureless objects and not even being able to find many of the galaxies on the charts. After several years of observing, I gradually found myself not only finding all the galaxies in the Tirion atlas but even glimpsing some too faint to be plotted, and galaxies that were formerly structureless now contained faint but definite spiral arms, knots, and dust lanes.

The best advice for someone who has not yet discovered the pleasures of galaxy viewing is simply to do a lot of looking. Don't be in a hurry to see lots of objects, but take plenty of time to look at each one and compare the view at many different magnifications. Let your eye relax and scan the galaxy, looking for subtle variations in the light. If you glimpse something you haven't seen before, keep looking and try to make certain of it. Once you have seen a detail, it will stick in your mind's eye and it will be easier to see the next time you observe.

A multitude of little things can improve your viewing, such as clean optics, a tube designed to eliminate stray light, and enhanced mirror coatings. Find the darkest site you can, even if it means a longer drive than usual, plan to observe when the dew points are low, and keep in mind that your dark adaptation will keep improving even after two or three hours.

Magnification

The last piece of advice may surprise those who have never tried it. Don't be afraid to magnify galaxies! Only the most diffuse objects are best at the lowest power. Almost all will be better seen at 15x per inch of aperture, and even higher power can be used on many.

Increasing the magnification helps in several ways. First, it makes the sky around the galaxy darker, which increases the contrast between the galaxy and the sky. More important, because the eye can not see detail that is both small and faint, magnification enlarges faint detail to the point where the eye can perceive it.

Because there are so many variable factors in observing, it is hard to predict in advance which powers will give the best view, so the best advice is simply to try different magnifications and compare the amount of detail you see at each level. When you reach the point where the overall image grows dimmer, look carefully and try even a little more power, because this is often the point where seeing faint features like spiral arms, knots, and dust lanes will be easier. Keep going and use your highest powers to examine the brightest parts of the galaxy like the nucleus and the prominent HII regions.

Numbers and Magnitudes

The table lists data for eighty-eight galaxies in Canes Venatici. Included are all those on the Tirion charts, plus about thirty more that are fainter than Tirion's cut-off magnitude. In the notes are some additional faint systems for which only magnitudes and relative coordinates are given.

The magnitudes of galaxies are problematic at best. The figures printed are from the most recently published sources, but keep in mind that many will appear fainter to your eye, especially face-on systems, dwarfs, and irregulars. Photographic magnitudes are marked with a "p" and tend to be fainter than what is seen at the eyepiece, so don't hesitate to search for a galaxy with a faint photographic magnitude. A 10-inch telescope will be able to show everything in Tirion plus a couple dozen more and should be able to reach at least some of the galaxies listed at magnitude 14.5p.

As long as this list is, it covers less than half the galaxies in *Uranometria 2000.0*. A 16-inch telescope should show most of these, but be forewarned that these charts include a number of dwarfs

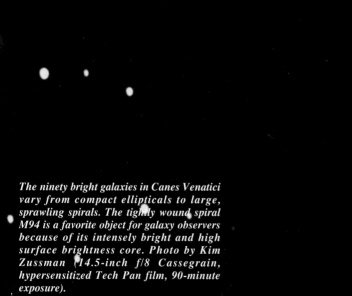

The ninety bright galaxies in Canes Venatici vary from compact ellipticals to large, sprawling spirals. The tightly wound spiral M94 is a favorite object for galaxy observers because of its intensely bright and high surface brightness core. Photo by Kim Zussman (14.5-inch f/8 Cassegrain, hypersensitized Tech Pan film, 90-minute exposure).

with very low surface brightness that will challenge even the most experienced eye.

Observations and Descriptions

The galaxies in Canes Venatici are so numerous that a small book would be needed to describe them all. The selected descriptions of a couple dozen of the best that follow are based on observations through a 10-inch f/6 scope plus a few views with a friend's 17.5-inch telescope and will proceed through the constellation in a clockwise direction.

NGC 4528, a posthumous addition to Messier's list as **M106,** is one of the ten brightest galaxies in the northern sky. It has an intense, small but nonstellar nucleus in a very large, bright, oval halo. The two spiral arms are subtle at first but can be seen as brighter parts of the halo on the northwest and southeast sides with the northern arm somewhat brighter than the southern one.

Forming a triangle with M106 are **NGC 4217,** 35' to the west-southwest, and **NGC 4220,** 45' to the north-northwest. NGC 4217, an edge-on spiral, is quite diffuse and is more difficult than the magnitude would suggest. The middle is larger than the ends, and the dust lane that appears in photos can not be seen in the 10-inch instrument, but readers with larger scopes can look for it. NGC 4220 is an easily seen bright oval. Of the five faint galaxies around these, NGC 4218 and NGC 4226 could just be glimpsed in the 10-inch telescope. A look through a 17.5-inch scope also showed the close pair **NGC 4231** and **NGC 4232,** plus **NGC 4248.**

NGC 4242 is a face-on spiral of low surface brightness. It is slightly oval with a little brightening in the middle, and it is diffuse and featureless even in a 17.5-inch telescope. Several bright points suspected as knots or the nucleus were assumed to be foreground stars because they remained stellar under high magnification. Later examination of a photograph showed many small knots but no foreground stars, so perhaps I really saw the knots.

NGC 4449 is one of the little-known treasures in this part of the sky. It is brighter than most of the galaxies in Messier's catalogue, and it takes magnification very well. It is an irregular galaxy seen face-on and its structure is similar to that of the Large Magellanic Cloud with a large central bar and only weak suggestions of arms. The bar is seen as a bright rectangular area and the brightest arm is the one on the northwest side and it gives the galaxy a somewhat of an "L" shape in the 10-inch telescope. Viewing at high magnification will show the nucleus and many knots, which are more numerous in the northern part. In large scopes, look for a dust cloud on the eastern side of the bar.

Near the border where Canes Venatici and Ursa Major meet are three easily seen ellipticals: **NGC 4111, NGC 4138,** and **NGC 4143.** The edge-on spiral NGC 4183 is more difficult. This galaxy is misidentified as NGC 4160 in Burnham's, Tirion, and the older *Atlas Coeli.*

NGC 4490 and **NGC 4485** form a splendid interacting pair only a few arcminutes apart, which will keep them in the same field of view even at high power. The major axes of the galaxies are approximately at right angles to each other. NGC 4490 is larger and brighter and has a prominent bulge on the end that is near NGC 4485.

Another pair that may be interacting, **NGC 4618** and **NGC 4625,** are a couple of degrees to the west and are about 9' apart. Both galaxies have similar redshifts and must be quite close to each other, but there is no detectable distortion of their shapes in a small scope.

If you usually pass up galaxy viewing because you have to observe from light-polluted areas, you will find **M94** (NGC 4736) worthwhile. It has a very intense nucleus surrounded by a bright, slightly oval glow. Visually it looks like an elliptical galaxy, because the dust that delineates the many spiral arms is not as dense as in many spirals. If you view from a dark site with a large scope, try to see a faint outer ring around the inner, brighter one.

NGC 4145 and **NGC 4151** are a pair less than a degree apart. NGC 4151 has a compact, bright central area in a very faint surrounding glow; NGC 4145 is visible as a large faint patch. In between there are three faint galaxies beyond the grasp of my 10-inch scope, but should be visible in a 16-inch scope. **NGC 4156** is on the northeast edge of NGC 4151, and the close pair NGC 4145A and UGC 7175 are 12' southeast of NGC 4145.

It is interesting to compare **NGC 4244** with the classic edge-on spiral NGC 4565 in Coma Berenices. NGC 4244 appears as a straight streak of light that is a little brighter in the middle, lacking both the central bulge and the prominent dust lane that so many edge-on spirals like NGC 4565 are famous for.

NGC 4214 is an irregular galaxy seen as a stellar nucleus in a oblong bright patch. As in many irregular galaxies, there is a bar with weak, knotty arms. In very dark skies, look for faint extensions on the ends of the brightest area. Observers with large instruments should also try high magnification to see if any knots are visible.

NGC 4395 has a similar structure but covers such a large area that it is quite a challenge to see. For this galaxy as with other extended objects of low surface brightness, the size of your telescope doesn't matter, because the darkness of the sky will determine visibility. Although the galaxy will still be very faint, a large scope will let you search for HII regions. There are quite a few of them, three of which on the southeast side received their own numbers in Dreyer's catalogue: NGC 4399, NGC 4400, and NGC 4401.

Does anyone at your star parties ever show off **NGC 4631**? If you have seen it, you know it is one of the most interesting galaxies in the entire sky and that too few deep-sky watchers are aware of this wonderful sight. Like NGC 4244, it is an edge-on spiral without a nuclear bulge or dust lane. Irregular dust patches and star clouds can be seen along its length as areas of varying brightness, especially at higher magnification. There are also irregularities in the outline, similar to but not as prominent as those in M82.

That NGC 4631 is thicker on the eastern end and not straight along its length is caused by its interaction with **NGC 4656** about half a degree to the southeast. There is a third galaxy in this group, the small elliptical **NGC 4627,** which can be seen just slightly northwest of the center of NGC 4631. Between the center of NGC 4631 and NGC 4627 and just off the edge of the larger galaxy is a foreground star.

NGC 4656 appears as a long streak with a turn of 90° on the eastern end. There are several bright patches in the curve and the brightest one on the end is **NGC 4657,** which for many years was thought to be another galaxy interacting with NGC 4656, but recently research has shown that NGC 4656 is one galaxy distorted by a recent encounter with NGC 4631.

Radio observations show that two neutral hydrogen arms between the galaxies and both galaxies have distorted shapes. Interestingly, the bright streak we see is only half the galaxy. The brightest patch on the southwest end is actually the nucleus with the other half of the galaxy stretching farther to the southwest. Try searching this area for an extension of the visible part. It is very faint on photographs and I have never heard of a visual sighting.

Among all the galaxies in Canes Venatici are two star clusters in our own Galaxy. The open cluster **Upgren 1** is a coarse grouping of seven 7- to 10-magnitude stars, five of which are true members of the cluster. What we see is the remnant core of a very old cluster; its other members have long since disappeared.

Another unusual thing about this cluster is that it lies so far from the Milky Way: it is only 10° from the north galactic pole, which means it is almost directly above us as the Sun and the other stars in the galactic disk orbit the galactic center. There are only three

Galaxies in Canes Venatici

Designation	Right Ascension	Declination (2000.0)	Size	Mag.	Type	Notes
NGC 4111	12h07.1m	+43°04'	4.8'x1.1'	10.8	S0	4109, mag 15.1p, 6' SSW
NGC 4117	12h07.8m	+43°08'	2.8'x1.1'	14.3p	S0	4118, mag 15.1p, 2' SE
NGC 4138	12h09.5m	+43°41'	2.9'x1.9'	12.3	E4	
NGC 4143	12h09.6m	+42°32'	2.9'x1.8'	12.1	E4	
NGC 4145	12h10.0m	+39°53'	5.8'x4.4'	11.0	Sc	Low surface brightness
NGC 4151	12h10.5m	+39°24'	5.9'x4.4'	10.4	P	
NGC 4156	12h10.8m	+39°28'	1.5'x1.3'	13.0	SBb	on NE edge of 4151
UGC 7175	12h10.9m	+39°45'	2.1'x0.6'	14.9	S	4145A, mag 15.5p, 1' W
NGC 4163	12h12.2m	+36°10'	1.9'x1.7'	13.0	Irr	4148, mag 14.6p, 30' SW
NGC 4183	12h13.3m	+43°42'	5.0'x0.9'	12.9	S	Mistaken for 4160 in Burnham and Tirion
NGC 4190	12h13.7m	+36°38'	1.7'x1.6'	13.3	Irr	
NGC 4214	12h15.6m	+36°20'	7.9'x6.3'	9.7	Irr	same as 4228; U727 mag 14.05, 25' SSW
NGC 4217	12h15.8m	+47°06'	5.5'x1.8'	11.9	Sb	4226, mag 14.4p, 7' SE
NGC 4220	12h16.2m	+47°53'	4.1'x1.5'	12.2	Sa	4218, mag 13.24, 15' N
NGC 4227	12h16.5m	+33°31'	1.8'x1.1'	13.8p	Sa	4229, mag 14.3p, 3' NE
NGC 4231	12h16.8m	+47°27'	1.4'x1.3'	14.5p	S0	4332, mag 14.6p, 1' S
NGC 4242	12h17.5m	+45°37'	4.8'x3.8'	11.0	S	Low surface brightness
NGC 4244	12h17.5m	+37°49'	16.2'x2.5'	10.2	S	
NGC 4248	12h17.8m	+47°25'	3.0'x1.2'	12.6	Irr	Low surface brightness
NGC 4258	12h19.0m	+47°18'	18.2'x7.9'	8.3	Sb	=M106
NGC 4288	12h20.6m	+46°17'	2.3'x1.7'	13.0	S	4288A, mag 15.0p, 3' S
NGC 4346	12h23.5m	+47°00'	3.5'x1.4'	12.2	E6	
NGC 4357	12h24.0m	+48°46'	3.8'x1.5'	13.5p	Sb0	
NGC 4369	12h24.6m	+39°23'	2.5'x2.4'	11.8	Sa	
NGC 4389	12h25.6m	+45°41'	2.7'x1.5'	12.5	SB	4392, mag 14.6p, 11' NNW
NGC 4395	12h25.8m	+33°33'	12.9'x11.0'	10.2	S	Very low surface brightness, contains H-II regions, 4399, 4400, and 4401
NGC 4449	12h28.2m	+44°06'	5.1'x3.7'	9.4	Irr	
NGC 4460	12h28.8m	+44°52'	4.4'x1.4'	12.3	SB0	
NGC 4485	12h30.5m	+41°42'	2.4'x1.7'	12.0	Irr	Interacting pair with 4490
NGC 4490	12h30.6m	+41°38'	5.9'x3.1'	9.8	Sc	Interacting pair with 4485
UGC 7699	12h32.8m	+37°37'	3.9'x1.2'	12.9	SBc	
NGC 4534	12h34.1m	+35°31'	3.0'x2.5'	12.3	Sd	Low surface brightness
Upgren 1	12h35.0m	+36°18'	15	7		Open Cluster
NGC 4618	12h41.5m	+41°09'	4.4'x3.8'	10.8	Sc	Possibly interacting with 4625
NGC 4619	12h41.7m	+35°04'	1.5'x1.5'	13.5p	SBb	
NGC 4625	12h41.9m	+41°16'	2.4'x2.0'	12.3	S	Possibly interacting with 4618
NGC 4627	12h42.0m	+32°34'	2.7'x2.0'	12.3	E4	Companion to 4631
NGC 4631	12h42.1m	+32°32'	15.1'x3.3'	9.3	Sc	Interacting with 4656
NGC 4656	12h44.0m	+32°10'	13.8'x3.3'	10.4	Sc	4657 is bright region on NE end; interacting with 4631
NGC 4736	12h50.9m	+41°07'	11.0'x9.1'	8.2	Sb	=M94
NGC 4800	12h54.6m	+46°32'	1.8'x1.4'	12.3	Sb	
NGC 4861	12h59.0m	+34°52'	4.1'x1.6'	12.2	Irr	Low surface brightness; a mag 13.2 HII region on SW end
NGC 4868	12h59.1m	+37°19'	1.7'x1.6'	13.1	Sb	
NGC 4914	13h00.7m	+37°19'	3.6'x2.2'	12.3	E2	
NGC 4956	13h05.0m	+35°11'	1.5'x1.5'	13.5p	S0	

Galaxies in Canes Venatici

Designation	Right Ascension	Declination (2000.0)	Size	Mag.	Type	Notes
IC 4182	13h05.8m	+37°36'	6.7'x5.9'	12.6	S	Low surface brightness
NGC 5005	13h10.9m	+37°03'	5.4'x2.7'	9.8	Sb	5002, mag 14.7p, 25' S
NGC 5014	13h11.5m	+36°17'	1.7'x0.7'	13.5	Sa	U8303, mag 13.22, 20' ESE
NGC 5023	13h12.2m	+44°02'	6.5'x1.0'	12.3	Sc	
NGC 5033	13h13.4m	+36°36'	10.5'x5.6'	10.1	Sb	
NGC 5055	13h15.8m	+42°02'	12.3'x7.6'	8.6	Sb	=M63
NGC 5074	13h18.4m	+31°28'	1.0'x1.0'	14.0	P	
NGC 5103	13h20.5m	+43°04'	1.5'x1.0'	13.6p	P	
NGC 5112	13h21.9m	+38°44'	3.9'x2.9'	11.9	Sc	5107, mag 13.66, 12' SSW
NGC 5123	13h23.2m	+43°04'	1.4'x1.2'	13.5p	Sc	5145, mag 12.98, 25' NE
NGC 5141	13h24.9m	+36°23'	1.7'x1.3'	12.8	S0	Low surface brightness; Markarian 451, mag 14.6p, 15' NW
NGC 5142	13h25.0m	+36°24'	1.1'x0.8'	13.3	S0	5143, mag 15.5p, 2' N; 5141, 2' SW
NGC 5149	13h25.9m	+35°57'	1.6'x0.9'	13.8p	Sb	5154, mag 14.9p, 6' NE
NGC 5173	13h28.4m	+46°36'	1.3'x1.3'	13.5	E0	5169, mag 14.47, 6' NNW
NGC 5194	13h29.9m	+47°12'	11.0'x7.8'	8.4	Sc	=M51; interacting with 5195
NGC 5195	13h30.0m	+47°16'	5.4'x4.3'	9.6	P	Interacting companion to 5194
NGC 5198	13h30.2m	+46°40'	2.1'x1.9'	12.7	E2	
NGC 5272	13h42.2m	+28°23'	16'	6.4		Globular Cluster = M3
NGC 5273	13h42.1m	+35°39'	3.1'x2.7'	11.6	E1	5276, mag 14.6p, 4' SE
NGC 5290	13h45.3m	+41°43'	3.7'x1.1'	12.6	Sb	5289, mag 13.5p, 13' SSW
NGC 5297	13h46.4m	+43°52'	5.6'x1.4'	12.2	Sb	5296, mag 14,70, 2' SSW
NGC 5301	13h46.4m	+46°06'	4.4'x1.1'	12.6	Sb	
NGC 5303	13h47.8m	+38°18'	1.1'x1.0'	12.0p	P	5303B, mag 15.3p, 4' S
NGC 5313	13h49.7m	+39°59'	1.9'x1.2'	12.6	Sb	5311, mag 13.6p, 10' W
NGC 5318	13h50.6m	+33°42'	1.5'x0.9'	13.5p	S0	5319, mag 15.0p, very close to E
NGC 5320	13h50.3m	+41°22'	3.5'x1.9'	12.7	Sc	
NGC 5326	13h50.8m	+39°34'	2.5'x1.3'	12.9	Sb	I4336, mag 14.6p, 12' N
NGC 5336	13h52.2m	+43°14'	1.6'x1.3'	13.6p	Sc	U8733, mag 13.60, 40' WNW
NGC 5337	13h52.4m	+39°41'	1.8'x0.8'	13.4p	S	5346, mag 14.9p, 10' SE
NGC 5347	13h53.3m	+33°29'	1.9'x1.5'	12.6	Sb	U8825, mag 14.4p, 7' NE
NGC 5350	13h53.4m	+40°22'	3.2'x2.6'	11.4	Sb	In a group of 5
NGC 5351	13h53.5m	+37°55'	3.1'x1.8'	12.1	Sb	5341, mag 14.1p, 12' SW; 5349, mag 15.1p, 4' SW
NGC 5353	13h53.5m	+40°17'	2.8'x1.5'	11.1	E5	5358, mag 14.6p, 5' E, both in a group of 5
NGC 5354	13h53.5m	+40°18'	2.3'x2.0'	11.5	S0	5355, mag 14.0p, 3' NE, both in a group of 5
NGC 5362	13h54.9m	+41°19'	2.4'x1.1'	13.1	Sb	
NGC 5371	13h55.7m	+40°28'	4.4'x3.6'	10.8	Sb	same as 5390
NGC 5375	13h56.8m	+29°10'	3.5'x3.0'	12.3	SBb	
NGC 5377	13h56.3m	+47°14'	4.6'x2.7'	11.2	Sa	
NGC 5380	13h56.9m	+37°37'	2.1'x2.1'	12.8	Sa	5378, mag 13.8p, 12' N
NGC 5383	13h57.1m	+41°51'	3.5'x3.1'	11.4	SBb	
NGC 5394	13h58.6m	+37°27'	1.9'x1.1'	13.0	SBb	Connected to 5395
NGC 5395	13h58.6m	+37°25'	3.1'x1.7'	11.6	Sb	Connected to 5394
NGC 5406	14h00.3m	+38°55'	2.1'x1.6'	13.0	Sb	5407, mag 14.5p, 15' NNE
NGC 5440	14h03.0m	+34°46'	3.3'x1.4'	13.4p	Sa	5441, mag 15.5p, 5' SSE
NGC 5444	14h03.4m	+35°08'	2.7'x2.3'	12.5	E1	5445, mag 14.1p, 7' S

Left: One of the oddest groups of galaxies in the sky is that made up of the edge-on spiral NGC 4631 (top) and NGC 4656, which lies about 30' southwest (bottom). These galaxies are gravitationally interacting, which causes a tidal distortion in both systems. The tiny elliptical NGC 4627 orbits NGC 4631, while the bright region on the northeastern end of NGC 4656 has the separate designation of NGC 4657. Photo by Martin C. Germano, who used an 8-inch f/5 Newtonian, hypersensitized Tech Pan, and a 50-minute exposure.

Top: The edge-on spiral NGC 4244 is only slightly brighter in its middle than at its edges and has a tiny stellar core. Photo by Martin C. Germano, who used an 8-inch f/5 reflector, hypersensitized Tech Pan film, and a 65-minute exposure.

Above center: One of the best examples of an irregular galaxy is NGC 4449, an object that appears square-shaped in a 6-inch telescope. Photo by Martin C. Germano, who used an 8-inch f/10 SCT, 103a-F film, and a 70-minute exposure.

Above: NGC 5005 is an oval-shaped galaxy measuring 5' across. Photo by Lee C. Coombs, who used a 10-inch f/5 Newtonian, 103a-O film, and a 15-minute exposure.

M64, the Blackeye Galaxy, is characterized by the broad patch of dust that obscures part of the galaxy's nuclear bulge. This feature is visible in a 6-inch telescope. Photo by Martin C. Germano, who used an 8-inch f/10 SCT, hypersensitized Tech Pan film and a 75-minute exposure.

Inset: The pair of interacting galaxies NGC 4485 (left) and NGC 4490 is visible in a 6-inch scope as a faint double patch of nebulosity. Photo by K. Alexander Brownlee, who used a 16-inch telescope.

Dozens of galaxies in Canes Venatici offer backyard observers detail when viewed on a dark night. One of the best of the constellation's galaxies, M106, is an 18'-long spiral that displays graceful spiral structure in 8-inch and larger scopes. The faint irregular galaxy NGC 4248 lies to the upper left. Photo by Chuck Vaughn, who used a 14-inch SCT at f/7, hypersensitized Tech Pan film, and a 60-minute exposure.

clusters closer to us than Upgren 1, two of which, the Coma Berenices cluster and the Ursa Major cluster, are in the same part of the sky.

About one hundred times farther away is the brilliant globular, **M3** (NGC 5272), only slightly fainter than the summertime favorites, M5 and M13. M3 is easily resolved in small scopes, and it is a stunning sight in the 10-inch scope at 150x. Although it is not as well known as some of the summertime globulars, M3 is one of the best globulars visible to northern observers.

NGC 4861 is another irregular galaxy with a structure like that of the Large Magellanic Cloud. The galaxy itself has a very low surface brightness and is hard to see, but on the southwest end there is a giant HII region that is easier to see than the rest of the galaxy. There is a foreground star on the northeast end. This galaxy also appears in the Index Catalogue as IC 3961, and you will sometimes see one number given to the galaxy and the other to the HII region, but the original description in Dreyer's catalogue fits the galaxy.

The numerous galaxies in the eastern part of Canes Venatici are more distant and will show detail in smaller telescopes. It is a rich area and one worth your while to explore, but detailed notes will have to wait for another time. A good place to start exploring this area is with the bright spiral **NGC 5371**. About half a degree to the west-southwest is a group of five galaxies within a span of 10'. The three brightest from north to south are **NGC 5350**, **NGC 5354**, and **NGC 5353**. Just to the east is **NGC 5355**, and to the southeast is the faintest, **NGC 5358**. The first four were easily seen in a 10-inch telescope, although the fifth may require a couple more inches of aperture.

NGC 5005 and **NGC 5033** are two nice spirals 40' apart. Both appear as ovals rapidly increasing in brightness to the middle. Three fainter systems are nearby. The brightest, **NGC 5014**, is 30' southwest of NGC 5033. The two fainter galaxies, **NGC 5002**, 25' south of NGC 5005, and **UGC 8303**, 25' south of 5033, will probably need a 12-inch telescope to be seen.

It is interesting to compare **M63** (NGC 5055) with M94 because both are many-armed spirals. M63 is more elongated because it is more highly inclined to our line of sight and the nucleus is less intense than M94. M63 also shows more distinct zones of decreasing brightness — at least three when the central area is viewed at high power. In both galaxies the dust lanes are not dense enough to appear as dark streaks, but look for areas surrounding the nucleus.

The reasons for the popularity of **M51** (NGC 5194) are not hard to understand. M51 is the perfect textbook illustration of spiral structure and one of the few galaxies where seeing the spiral pattern is easy at the eyepiece. In addition it has a bright interacting companion apparently connected to the end of one of the arms. What is the smallest aperture needed to see the arms? Although some brightness variations may be seen in smaller scopes, I would say 8 inches are necessary to really follow them for any length and then only in very good conditions. The brightest arm is the one between the nucleus of the larger galaxy and the companion. If you see this try following it around and then try spotting the other arm that goes out to the companion. When looking for the bridge, try to be honest about whether you truly see it or are remembering it from photographs. With a 10-inch telescope I have not been sure I have seen it, because it is a subtle feature clearly seen only in larger instruments. As with most galaxy features, the spiral arms are easiest to see when the galaxy is magnified to the point where it appears dimmer.

After viewing M51, you might want to try for three faint galaxies nearby. NGC 5198 is half a degree south of M51, and 20' west of it is the pair 5169 and 5173. All are small and faint but within the grasp of a 10-inch scope.

Beyond the galaxies described, there are several dozen more that can be seen with 8- to 10-inch scopes, and this only scratches the surface of the possibilities of an instrument with twice the aperture. Good luck as you explore this fascinating area of the sky.

Max Radloff lives in South St. Paul, Minnesota. A musician by profession, he revived his childhood interest in astronomy about ten years ago. He is secretary of the Minnesota Astronomical Society and writes a deep-sky column for its newsletter, "Gemini."

The field of NGC 7331 is shown here in this photo by K.A. Brownlee. The three galaxies to the east of 7331 are NGC 7335, NGC 7336, and NGC 7337. All other noted objects are stars or close double stars.

NGC 7331

And Its Ambiguous Galaxies

by Jeffrey Corder and Steven Gottlieb

The brightest galaxy in Pegasus, NGC 7331, lies high overhead for northern viewers on fall evenings. This spiral was one of William Herschel's earliest discoveries over 200 years ago. Since then, it has been carefully studied by astronomers using progressively more sophisticated instruments and today is considered a galaxy much like our own Milky Way.

Classed as "spiral nebulae" by early observers, galaxies, they noticed, are often accompanied by smaller, fainter neighbors. NGC 7331 is no exception. Following Herschel's discovery, various observers noted the presence of eight suspected companions. Modern observations support the likelihood of a true galaxy cluster in the background, but NGC 7331 is probably not involved. Chance alignment has the galaxy in the foreground along the same line-of-sight as the small, more distant group.

Discoveries of new galaxy companions were often a pathway to fame and success for late 18th- and early 19th-century astronomers. Monstrous-size telescopes employed by these observers were able to reveal innumerable small "nebulae," but were cumbersome to use, making the note-taking process difficult and drawn-out. Errors were inevitable. Moreover, the telescopes were often situated in locations subject to extended spells of bad weather. Astronomers therefore found it difficult to recheck their notes and observations. The NGC 7331 group fell victim to these errors, some of which have been passed down to more modern times.

Although visible in large finder scopes, NGC 7331 was apparently missed by Messier. In many respects this galaxy resembles the famous Andromeda spiral. Like M31, the Pegasus galaxy is much elongated. This appearance is caused by its nearly edge-on alignment. It has been determined that the western side is nearer to us, hence the galaxy is believed to be rotating with the spiral arms trailing. Lying 23 times more distant than M31, and with an angular size one-twentieth that of the nearer galaxy, it can be seen that NGC 7331 is comparable in size. Also, like M31, 7331 is rich in HII regions; over 125 have been catalogued. These clumps of ionized hydrogen allow 7331's mass, rotation, and distance to be measured with accuracy. The main

Table One

NGC	DISCOVERER	R.A.	DEC.	NOTES	REF.
7325	Schultz 1882	22h 35.7m	+34°15'	"F,vS,7331 foll."	1,3,4
7326	Rosse, 1874	22h 35.7m	+34°18'	"eF,eS,7331 foll."	3,1
7327	Tempel	22h 35.7m	+34°21'	"eF,eS,Np 7331"	5,1
7331	W. Herschel	22h 35.9m	+34°18'	"B,pL,pmE 163°,smBm"	1,3,4
7333	Schultz, 1882	22h 36.1m	+34°19'	"VF,VS,p 7335"	1,3,4
7335	W. Herschel	22h 36.1m	+34°19'	"VF,VS,pos 61°,dist 228''"	1,3,4
7336	Rosse, 1849	22h 36.1m	+34°21'	"eF,VS,pos 43°,dist 315''"	1,3,4
7337	Rosse, 1849	22h 36.3m	+34°15'	"eF,S,Stell,p119°d328''"	1,3,4
7338	Tempel	22h 36.3m	+34°18'	"eF,eS,Sf 7335"	5,1
7340	Rosse, 1849	22h 36.5m	+34°18'	"VF,VS,pos 92°,dist 507''"	1,3,4

Bibliography

1. Dreyer, J.L. *New General Catalogue of Nebulae and Clusters of Stars*. London: Royal Astronomical Society, 1888.
2. Sulentic, J. and Tifft, W. *The Revised New General Catalogue of Nonstellar Objects*. Tucson: University of Arizona Press, 1973.
3. Rosse, William Parsons, Lord. *Observations of Nebulae and Clusters of Stars Made with the Six-foot and Three-foot Reflectors at Birr Castle from the Year 1848 up to the Year 1878*. Dublin: Royal Dublin Society, 1880.
4. Schultz, Dr. Herman. *Monthly Notices of the Astronomical Society* 35 (1874-75): 135.
5. Tempel, W. *Astronomische Nachrichten* 102-103, no. 2439 (1882).

difference between M31 and 7331 is the greater abundance of younger, bluer stars in the Pegasus galaxy.

Recently, sources of radio emission have been observed in the region of NGC 7331. The sources also seem to engulf the nearby galaxy cluster, Stephan's Quintet, located about 1.5° to the southwest. This, along with the photographic evidence of a very faint but extensive filament of material stretching between Stephan's Quintet and NGC 7331, argues for a past gravitational interaction that may have once occurred between these galaxies. NGC 7320, the largest member of the Quintet, lies in the foreground of its more distant members, and is the most likely candidate for an ancient association with NGC 7331.

Observers with small telescopes can easily view the prominent features of NGC 7331. It has a well-defined oblong core, and a bright, compact nucleus. The north-south orientation is also quite plain. Spiral structure is evident in 12-inch and larger telescopes. Remarkably, some of the associated HII regions can be seen in larger instruments, especially of the 16-inch class. Careful observers with large telescopes, keen eyesight, and dark skies may catch glimpses of the string of HII knots aligned along the western edge of the galaxy. As with M31, a dust lane is visible.

As noted in the first table, a total of nine NGC entries lie in the vicinity, including NGC 7331. A second table includes five questionable NGC entries, with their RNGC coordinates and descriptions. Again, although the galaxies listed in the tables seem to be companions of NGC 7331, even a casual inspection of photographs reveals NGC 7331 to be a foreground object.

Observers using 8-inch telescopes can see only a couple of the galaxies listed in the first table, partly due to their dimness, but also because of the many ambiguous galaxies catalogued in this region. Let's explore the entries for the NGC 7331 region in historical detail and find the true members.

A comparison of the two tables reveals a discrepancy in positional data for the first two entries, NGC 7325 and NGC 7326. Both objects were studied by Lord Rosse and other Irish observers at Birr Castle in 1874. Accurate

K. Alexander Brownlee

positions were measured in reference to NGC 7331 and it is clear that the original observers mistook the slightly elongated appearance of close double stars for very small galaxies. The authors of the RNGC evidently used these two ambiguous NGC designations to incorrectly label two nearby anonymous systems, both 10' northwest of the original NGC positions.

In actuality, NGC 7326 is a close pair of magnitude 15 stars just west of the core of NGC 7331, and NGC 7325 is a wider double easily resolvable in a 10-inch scope, 3' south. It is likely that poor resolution of the older telescopes, compared to even the common smaller compound telescopes of today, misled the earlier astronomers into making such false discoveries as these two "galaxies." The newer RNGC 7326 is listed in Fritz Zwicky's *Catalog of Galaxies and Clusters of Galaxies* (CGCG) as 514.066 at photographic magnitude 15.7 and is visible in backyard scopes.

The next object, NGC 7327, was discovered by the German astronomer Ernst W. Tempel with an 11-inch refractor. Although Tempel gave only a vague position in his original paper, Dreyer gave an exact position in the NGC that closely matches Tempel's description of "preceding the northern end of NGC 7331." This information was undoubtedly conveyed through private correspondence. No double star is noted at the exact position from the NGC, but a single faint star lies very near the expected position. Astronomer Harold Corwin, Jr. of the University of Texas at Austin believes that some of the nebulosity involved with NGC 7331 may have caused the star to appear nonstellar to early observers, but he is guarded in his opinion. A visual inspection of the star in question shows not the slightest hint of nebulosity. It is likely that Tempel was misled during a period of poor seeing. Since Tempel claims to have seen all nine "nebulae" in this region it is also assumed this could not be a duplicate of another of his observations. We may never know exactly which object Tempel actually saw, but it is plain that no galaxy exists at the given NGC position.

Our next object, NGC 7333, was discovered by Dr. Herman Schultz at Uppsala, Sweden and announced in January 1875. Fortunately, Dr. Schultz gave a precise position for his new "nova" to the east of NGC 7331. The astronomers at Birr Castle, using their tremendous 72-inch reflector did not find the Schultz object, which right away cast doubt on its existence. Modern telescopes show an extremely close double star at that location, which is just elongated on a Hale 5-meter plate. Amazingly, Schultz must have noted this elongation visually in his 13-foot focal length refractor. The exact match in position leaves little doubt as to the certainty of the identification. The RNGC correctly notes this as a double star.

NGC 7335 is one of the galaxies that does indeed exist. Discovered by William Herschel, this galaxy is the brightest of the group of four on the eastern side of NGC 7331. It lies in Herschel's class III, "very faint nebulae." Because of its great distance only photographs reveal much detail, but observers will note its distinctly enlongated shape and large bright core.

The other three galaxies positioned on the eastern side of NGC 7331 can be seen in 10-inch or larger telescopes. One of these is a nearby companion to NGC 7335, the barred spiral NGC 7336 only 2' north-northeast. This system was discovered by Lord Rosse but seems to

Table Two

RNGC	R.A.	DEC.	RNGC DESCRIPTION	CONCLUSION
7325	22h 35.4m	+34°22'	"R,alm stel,HISB	not NGC object,10'NW
7326	22h 35.3m	+34°25'	"EL,sldiff,BM,sstsusp	not NGC object,10'NW
7327	22h 35.7m	+34°21'	"*"-Carlson	star NW 7331
7333	22h 36.1m	+34°19'	"**"-Carlson	close double NNW 7331
7338	22h 36.3m	+34°18'	"**"-Carlson	double star NNE 7337

Bibliography

1. Sulentic, J., and Tifft, W. *The Revised New General Catalogue of Nonstellar Astronomical Objects*. Tucson: University of Arizona Press, 1973.
2. National Geographic Society, "Palomar Observatory Sky Survey." (print no. 0-778). Pasadena, California: California Institute of Technology, 1954.
3. Carlson, Dorothy. "Some Corrections to Dreyer's Catalogue of Nebulae and Clusters." *Astrophysical Journal* 91 (1940): 350.
4. *Publications of the Lick Observatory. Vol. 3*. Sacramento: University of California Publications, 1918.

Also: private correspondence with Dr. Harold Corwin, Jr., The University of Texas at Austin, Astronomy Dept. RLM 15.308. Austin, TX 78712, 12 Sept. 1985.

---

have been missed by Zwicky in the CGCG. Otherwise, all of the early observers seem to have their identifications correct in this case. Visually NGC 7336 is very faint and elongated in a northeast-to-southwest direction. It appears very small and contains a bright core. Although the faintest of four galaxies in this group, it should present no challenge to a 17.5-inch telescope.

The third small galaxy that can be viewed in backyard telescopes is NGC 7337, again discovered by Lord Rosse. This barred-spiral system is located about 7' south of the NGC 7335/36 duo. A 13th-magnitude star appears nearly attached on its southeast side. The galaxy is very faint, very small, and round, and has an even surface brightness. Brighter NGC 7340 is just to the northeast.

The second of Temple's two "discoveries" in this area has been catalogued as NGC 7338. Only vaguely referred to in Tempel's original paper, it is noted as being "between the four brighter nebulae to the east, a little closer to the two southern ones." We can only assume that he was able to satisfy Dreyer with a more precise location, than which appears in the NGC. This time the NGC position does not fall on any visible star, double star, or galaxy. Corwin believes the best evidence points to a faint double star 3' southeast of NGC 7335. This position differs with the NGC position by only 0.2' of right ascension. The NGC position may be off by this much, as it seems then to fit Tempel's original notes more accurately. At any rate, a visual inspection of the star in question shows it to be double, which could be mistaken for a nearly stellar galaxy.

The final galaxy for consideration in the NGC 7331 area is NGC 7340, a faint elongated system 8' to the east of NGC 7331. The galaxy was the third and final discovery by Lord Rosse at Birr Castle, Ireland, in 1849. This galaxy is considerably faint, at only photographic magnitude 14.9, which places it as second brightest among its three faint companions. It appears round with a small bright core. This galaxy is at the very limits of an 8-inch telescope under a dark sky.

From our considerations of the field surrounding NGC 7331, it is obvious that early astronomers realized that these "nebulae" often appear in groups. Observers evidently searched the area around the brighter galaxies quite carefully, hoping to make a discovery of their own. A good example of this method is illustrated in the catalogue of the astronomers using the 72-inch telescope at Birr Castle. Even Dreyer noted that the Birr Castle "discoveries" were usually found clustered together around brighter galaxies. In our case, in the vicinity of NGC 7331, early observers correctly identified four neighbors (NGC 7335, 7336, 7337, and 7340), but mistakenly observed five others.

It is also evident from this one case that many errors still exist in the catalogues used regularly by amateurs. It is important to spend some time to research observations so that we can be certain of what we are seeing and recording. This requires some effort but it can make the difference between accurate and inaccurate observations.

Jeffrey Corder is a deep-sky observer living near Fort Meyers, Florida. His last article, "Observing the IC 4329 Galaxy Group," appeared in Deep Sky *#14. Steven Gottlieb is an avid deep-sky observer living in El Cerrito, California.*

Small Scope Showcase
Dwarf Galaxies for "Dwarf" Telescopes
by Alister Ling

The spring nights almost seem balmy in comparison to cold winter observing sessions. With the air at a few degrees above freezing, the gloves can come off once more. I always marvel at the simple reaction of the body to habit. In autumn, similar temperatures bring on a deep chill.

In a previous column I made the suggestion that small scopes can do useful and rewarding work supernova hunting. This is true even when the general transparency conditions are not very good. The main idea here is that the parent galaxy may be nearly invisible, but field stars and supernovae can shine through the light pollution and haze at medium to high powers.

What about the other extreme? A small scope functioning at low power can pick up very low surface brightness galaxies. E. E. Barnard taught us this when he made his discovery of NGC 6822 in Sagittarius with a 5-inch refractor back in 1884. Dwarf galaxies are quite the challenge for "dwarf" scopes. Instead of looking for spiral arms or dust lanes, we are aiming for simple detection. Over the years we have learned that we can experience quite a thrill from these feats of observational ability.

M33 (NGC 598) in Triangulum is the prototype large low surface brightness (LSB) galaxy. Readily visible in binoculars, it can get lost when you are trying to locate it with a decent aperture under less than ideal conditions. With this galaxy, and the next few which follow, contrast is the key parameter instead of aperture.

It pretty well goes without saying that the sky condition is the major factor in determining the visibility of LSB galaxies. Slight cirrus, haze, and aurora can kill the possibility. Before discounting any galaxy as invisible, you should give it a try on several occasions. You may with to wait until solar minimum to ensure the weakest auroral conditions if you live in the north.

Short refractors are definitely the best telescopes for LSB galaxy hunting. Their insides are well baffled, and together with their typically clean objective lenses keep scattered light to a minimum. This is critical for an object like the dwarf **Leo I**, just 20' north of Regulus. An open framework Newtonian has a tough time here since Regulus

Leo I is an extremely low surface brightness galaxy just north of the bright star Regulus. Photo by Martin C. Germano (5.5-inch f/1.65 Schmidt camera, hypered Tech Pan film, 5-minute exposure).

Left: NGC 2403 is a bright spiral that is easily visible in small scopes. Photo by Martin C. Germano (8-inch f/5 reflector, hypered Tech Pan film, 60-minute exposure).

scatters light off dust, tube, and focuser, to name but a few sites. The trick is of course having a refractor big enough, but the 10' by 8' diffuse glow might well be within the grasp of a 5-incher. There simply aren't enough refractor owners who attempt deep-sky challenges for us to know.

If Leo I is too much of a challenge, try **NGC 2403** or **IC 342** in Camelopardalis. Both of these galaxies have been reported in scopes as small as 6 inches. Your greatest challenge will be aiming your scope at the correct spot in space (*Sky Atlas 2000.0* is a great help). These two objects are well off the beaten path, but a nice sight to behold.

The Herschels can be forgiven for having missed IC 342. The sweeping power of the 18-inch reflector was 157x, giving a field of view of 15'. At a size of 18' by 17', IC 342 would have flowed out of the field! It certainly doesn't help that it is another magnitude fainter than NGC 2403.

No matter what the size of your scope, you should try to observe some H II regions in IC 342. These giant M42s or M17s might respond well to an O III filter. The galaxy will effectively disappear from sight, but perhaps a diffuse faint knot or two will remain visible.

Much easier to locate, but not necessarily as easy to see, is **NGC 4236**. This galaxy sits between the two tail stars of Draco, Kappa and Iota. NGC 4236 is a biggie, too — 19' by 7' and oriented almost north/south.

One of the observing techniques best suited for LSB objects is sweeping or rocking the scope back and forth over the charted position. The dark adapted eye is remarkable sensitive to the motion of subtle contrasts. At first all you may detect is that the background sky is somewhat brighter than the surrounding area. As your eyes dark adapt further, you may notice that the glow of NGC 4236 is not perfectly uniform.

The greatest challenge of this column's galaxies is the **Draco dwarf gal-**

NGC 4236 is one of the prettiest galaxies in the sky. Photo by Martin C. Germano (8-inch f/5 reflector, hypered Tech Pan film, 75-minute exposure).

Left: Face-on IC 342 is a low surface brightness challenge. Photo by Martin C. Germano (8-inch f/5 reflector, hypered Tech Pan film, 45-minute exposure).

axy. It can be found on *Uranometria* chart #52 not too far from the head of the dragon. It is a whopping 33' by 19' in extent, bigger than the Helical Nebula. Sweeping will be your best bet to detect this dwarf, coupled with properly adapted eyes that have not seen even a red flashlight for several minutes. The standard dark cloth hood will help even more. Don't give up too easily!

A big problem that often frustrates spring galaxy observers is dew. It especially strikes hard at the small scope owners. Although all telescope optics radiate thermal energy to space more or less equally, smaller optics contain less mass and therefore cool off faster. The nights are frequently loaded with moisture, which is quick to condense on lenses, mirrors, secondaries, and of course charts.

Dew does two things to the image. The most obvious effect is to create halos around the brighter stars, followed by the fainter ones as the condensation sets in. The second effect is to make everything fainter, not only by reducing the reflectivity, but also by scattering the available light around the field of view. Scattering is what creates the halos in the first place.

The gradual deterioration of the image can go unnoticed in spring because of the lack of bright field stars in galaxy territory to ring the alarm. Every galaxy begins to appear undefined at the edges. The light drop off at the edges of many galaxies is quite fast, but the scattering effect of the dew gives the edges, and especially the near-stellar cores of the galaxies, a rather "soft" appearance.

Further, the observer only gradually becomes aware that the magnitudes given seem a bit off — surely those 12th-magnitude objects are another unit fainter! The only way to prevent the night from being a short one is to set up heaters or keep a hair dryer handy. I've never liked hair dryers for two reasons. One is that they might spit some dirt specks out at the optics. But my main dislike is that I usually have to lose some dark adaptation with a brighter red flashlight to check if the dew is gone. This can be really frustrating if you are trying to push deep in galaxy clusters: After you've dried up the optics, the brighter galaxies are at first harder to see! If your scope is fairly small, you can't allow yourself the luxury of losing a magnitude.

Let's face it, we all struggle against the weather and time to get in some good observing, so we owe it to ourselves to use a couple of cloudy nights to wire up some heating units to make the spring nights as long as possible. Whether you're into marathoning for the Messier objects or excavating LSB galaxies, you'll need to be operational all night long!

Alister Ling is a Canadian deep-sky observer with wide expertise in nebular filters, planetary nebulae, galaxies, clusters, and small telescope observing.

A Trio of Springtime Galaxy Groups

by Jeff Corder

No two clusters of galaxies are exactly alike. Each clump of spirals and ellipticals in our night sky has its own individual character, its own unique makeup. Some clusters like Abell 2151 in Hercules are rich in spirals, while others contain relatively few spirals but many ellipticals, like the Perseus Galaxy Cluster, Abell 426. Others are classified as cD groups because they are dominated by a central cD galaxy. Such is the case with Abell 2199, the brightest member of which is the enormous cD galaxy NGC 6166. Still other galaxy groups, like the Virgo Cluster, appear to have an even distribution of members.

The accepted system relating galaxy groups is the Rood-Sastry classificatin scheme, developed by Herbert J. Rood of Princeton University and K. L. V. N. Sastry of Canada's University of New Brunswick. The Rood-Sastry system classifies galaxy clusters based on their visual and photographic appearances and differences in the distribution of indivudual members.

Fortunately springtime is the best time for viewing galaxy groups and clusters. Coma Berenices, Leo, and Virgo each contain heavy concentrations of galaxies. The Virgo Cluster, the most observed concentration of galaxies, is one of the brightest galaxy groups. However, its members are sparsely scattered over a large area, which make it appear not so much as a cluster but as a wide field of small galaxies covering many degrees of sky. Compare it with the region's rich clusters like the Coma Cluster or Abell 1367.

A few of springtime's galaxy groups are worth a detailed look. With today's large-aperture telescopes these groups hardly present a challenge to persistent observers. They do present new ground, unconquered territory for those who want to explore the limits of the galaxies they can observe. Three of these galaxy groups represent a particularly meaningful range over the Rood-Sastry scheme and offer plentiful challenges to backyard observers.

The **NGC 4169 group** contains only four bright members, all of which are visible in 8-inch or larger telescopes. To find this galaxy group, aim your telescope 3° west-northwest of Gamma Comae Berenices. You'll see a 6th-magnitude star accompanied by an 8th-magnitude star 10' to the northeast. (You may also see the faint galaxy NGC 4185 just south of the fainter star.) Place the cross hairs of your finder scope midway between these stars and slowly move the telescope 40' north. You are now centered on the NGC 4169 galaxy group.

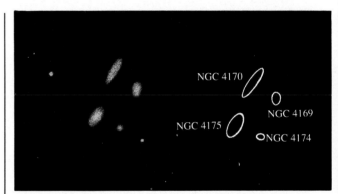

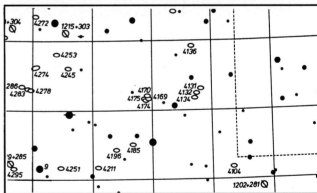

Top: This sketch of the NGC 4169 group has north at top and east at left; galaxies are identified at right. Jeff Corder recorded the group with a 12.5-inch f/6 reflector. Above: A portion of Uranometria 2000.0 *shows the position of the NGC 4169 group. Copyright 1987 Willmann-Bell, Inc.*

The NGC 4169 Galaxy Group

Galaxy	R.A.(2000.0)Dec.	Mag.	Size
NGC 4169	12h12.3m +29°11'	12.9	2.1' by 1.1'
NGC 4170	12h12.4m +29°13'	13.5?	—
NGC 4174	12h12.5m +29°09'	14.3	0.8' by 0.3'
NGC 4175	12h12.5m +29°10'	14.2	2.0' by 0.5'

A Slice of the Sky

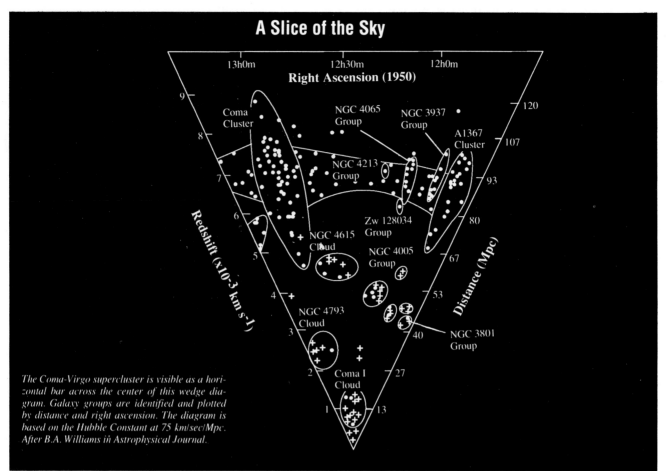

The Coma-Virgo supercluster is visible as a horizontal bar across the center of this wedge diagram. Galaxy groups are identified and plotted by distance and right ascension. The diagram is based on the Hubble Constant at 75 km/sec/Mpc. After B.A. Williams in Astrophysical Journal.

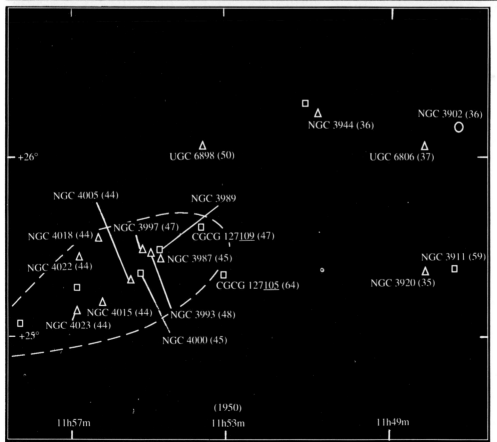

Magnitudes, identities, and radial velocities are shown for the NGC 4005 group in this diagram after B.A. Williams. The outline of the cluster is indicated by a dashed line.

The Rood-Sastry Scheme (simplified)

cD. **Central-dominant.** Cluster is dominated by a single giant galaxy, often an elliptical, surrounded by a halo of dwarf galaxies.

B. **Binary.** Cluster is dominated by a pair of giant galaxies, usually ellipticals, surrounded by a halo of dwarf galaxies.

C. **Core-rich.** Cluster has a core of three or four large galaxies that are much brighter than other members.

L. **Line.** Dominant galaxies in the cluster are oriented in a chain or line (usually three or four galaxies).

F. **Flattened.** The cluster's few brightest members are evenly arranged and somewhat flattened.

I. **Irregular.** Cluster has an ill-defined symmetry, or its ten brightest members are randomly dispersed.

The galaxies in this group are arranged in a rectangular shape. **NGC 4169**, a lenticular galaxy with a photographic magnitude of 12.9, marks the rectangle's northwest corner. Although this galaxy's dimensions are 2.1' by 1.1', when I viewed it with a 12.5-inch telescope, NGC 4169 appeared to be a slightly asymmetrical fuzzball measuring 50" by 40" (oriented in p.a. 156°) with a bright nucleus. The rectangle's northeastern marker is the surprisingly faint galaxy **NGC 4170**. It is an edge-on galaxy with a low surface brightness. In my 12.5-inch scope, NGC 4170 measures 80" by 35" (in p.a. 160°) and contains a slightly brighter core.

The southern half of the rectangle is made up of **NGC 4174**, an extremely small galaxy with a stellar nucleus, and **NGC 4175**, a lens-shaped galaxy. NGC 4174 is an object of unknown morphological type that glows dimly at photographic magnitude 14.9 and has an overall size of 0.8' by 0.3'. NGC 4175 is a spiral galaxy measuring 2.0' by 0.5' and glowing at photographic magnitude 14.2. Visually NGC 4175 measures 60" by 35" in p.a. 155°.

After you finish inspecting the four galaxies in the NGC 4169 group, swing your telescope 5° southwest. Here you'll find a rich clump of faint galaxies clustered around a solitary 8th-magnitude star. This is the **NGC 4005 galaxy group**. You'll be amazed at the contrast between this group and the previous one. The NGC 4005 group is much larger and richer, containing fourteen NGC galaxies in a diameter of 40'. The few brightest members can be glimpsed with an 8-inch telescope, but experience dictates that a 16-inch or 17.5-inch scope is necessary to spot the faintest ones.

The northwest end of the group contains the fine edge-on spiral galaxy **NGC 3987**. This object is the largest in the group, measuring 2.5' by 0.5', and glows at photographic magnitude 14.4. Tiny **NGC 3989**, a magnitude-15.7 galaxy measuring only 10" across, lies just east of NGC 3987.

Several arcminutes farther east lie the easier targets **NGC 3993** and **NGC 3997**. These

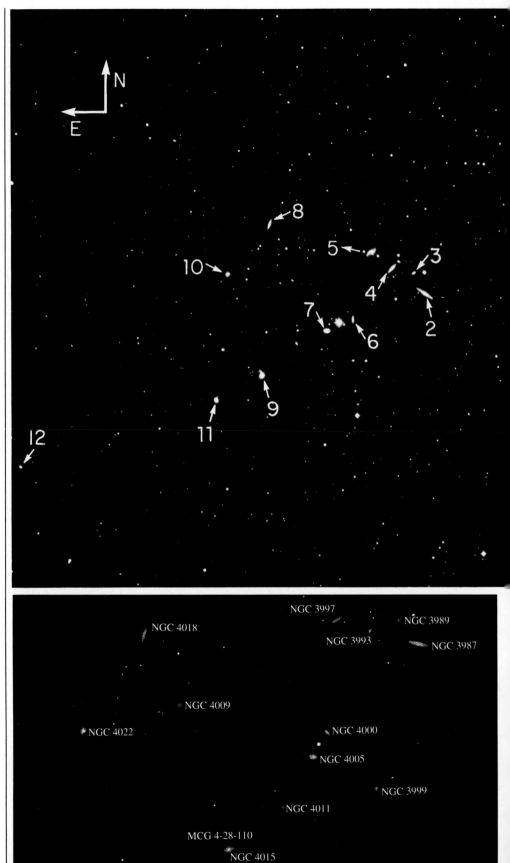

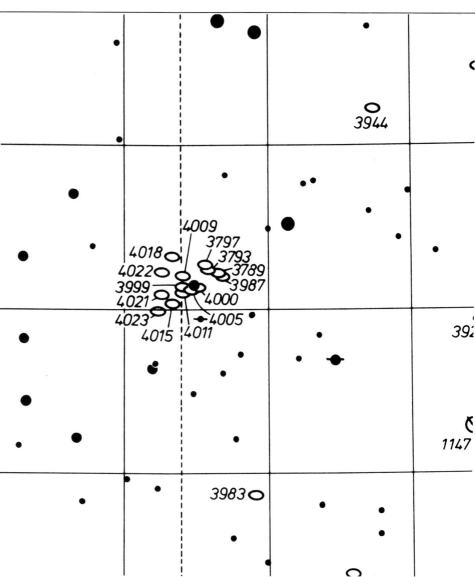

Left: The NGC 4005 group sketched by Jeff Corder using a 17.5-inch f/4.5 telescope. North is approximately at top and east to the left. Above: This photo of the NGC 4005 group is adapted from the blue plate of the Palomar Observatory Sky Survey. Labeled galaxies are: Zwicky 127-109 (1), NGC 3987 (2), NGC 3989 (3), NGC 3993 (4), NGC 3997 (5), NGC 4000 (6), NGC 4005 (7), NGC 4018 (8), NGC 4015A and NGC 4015B (9), NGC 4022 (10), NGC 4023 (11), and Zwicky 127-133 (12). From B.A. Williams, Astrophysical Journal. Above right: A finder chart from Uranometria 2000.0 shows the cluster's bright members. Copyright 1987 Willmann-Bell, Inc.

Bright Members of the NGC 4005 Galaxy Group

Galaxy	R.A.(2000.0)	Dec.	Mag.	Type	Size
NGC 3987	11h57.1m	25°11.6'	14.4	Sbc	2.3' by 0.4'
NGC 3989	11h57.1m	25°13.7'	15.7	Sbc	0.7' by 0.3'
NGC 3993	11h57.5m	25°14.4'	14.8	Sbc	1.6' by 0.4'
NGC 3997	11h57.6m	25°16.3'	14.3	S(B)bc	1.6' by 1.3'
NGC 3999	11h57.7m	25°05.3'	15.7	—	—
NGC 4000	11h57.7m	25°08.5'	15.2	Sc	1.1' by 0.2'
NGC 4005	11h57.9m	25°07.3'	14.1	Sb	1.1' by 0.6'
NGC 4009	11h58.1m	25°12.3'	—	—	—
NGC 4011	11h58.1m	25°05.6'	15.7	—	—
NGC 4015	11h58.7m	25°02.3'	14.2	—	0.8' by 0.2'
MCG428110	11h58.7m	25°02.3'	15.7	Pec	—
NGC 4018	11h58.7m	25°18.8'	14.7	Sc	1.8' by 0.3'
NGC 4021	11h58.8m	25°04.7'	15.3	—	—
NGC 4022	11h58.8m	25°13.3'	14.4	S0	1.3' by 1.2'
NGC 4023	11h58.8m	24°59.3'	14.6	Sb	1.1' by 0.7'

Above: Jeff Corder sketched the NGC 3937 galaxy group using a 17.5-inch f/4.5 reflector. The finder chart, from Uranometria 2000.0, shows major galaxies in the cluster. Copyright 1987 Willmann-Bell, Inc.

Fundamental Data for the NGC 3937 Galaxy Group

Galaxy	R.A. (2000.0) Dec.		Mag.	Type
#67	11h50.6m	20°54.4'	15.5	E1
#68	11h50.7m	21°09.7'	15.3	—
#70	11h50.7m	20°29.6'	15.4	E1
#71	11h50.7m	21°08.5'	15.4	—
#72	11h50.8m	20°23.7'	14.6	S pec
IC 742	11h50.8m	20°47.7'	15.1	—
#74	11h50.9m	20°59.7'	15.0	E1
NGC 3929	11h51.7m	20°59.8'	14.5	E1
#82	11h51.7m	21°06.5'	14.7	S pec
#83	11h52.1m	21°05.7'	15.1	—
IC 2968	11h52.5m	20°37.4'	15.5	—
NGC 3937	11h52.7m	20°37.6'	14.0	E1
NGC 3940	11h52.7m	20°59.3'	14.3	E1
NGC 3943	11h52.7m	20°28.5'	14.7	—
#92	11h52.9m	20°39.4'	15.3	E1
NGC 3947	11h53.1m	20°44.8'	14.2	S pec
NGC 3946	11h53.1m	21°01.3'	15.5	—
NGC 3954	11h53.7m	20°52.7'	14.4	E1
#100	11h53.9m	20°34.3'	14.9	—

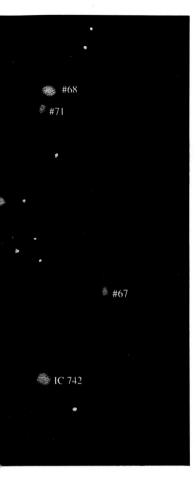

galaxies are nearly twins: both appear to be nearly edge-on and have a lenticular shape. NGC 3993 is an Sb-type spiral glowing at photographic magnitude 14.8 and spanning some 1.9' by 0.6'. In the 12.5-inch scope this galaxy appears to measure 75" by 20" in p.a. 110°, and it shows a very faint nucleus. NGC 3997 is a peculiar, barred spiral with a photographic magnitude of 14.3 and a diameter of 1.8' by 1.0'. Visually this object can be seen lying between two faint stars and measures 80" by 25" in p.a. 100°. Some 15' east lies another edge-on spiral, **NGC 4018**. This object also has a lenticular appearance and is relatively faint with the 12.5-inch scope. It measures 100" by 25" in p.a. 115°.

As you move southward, **NGC 4022** pops into view. NGC 4022 is a face-on lenticular galaxy that dimly glows at photographic magnitude 14.4 and spans 30". The most noticeable observational characteristic of NGC 4022 is its prominent condensed nucleus. Extremely faint **NGC 4009** can be glimpsed 10' to the west. This one requires a sharp eye and clean optics, and even then averted vision will show it best. It appears slightly oval and seems to be about 20" in diameter.

The brightest galaxy in this cluster is **NGC 4005** itself, an Sb-type system lying near the central 8th-magnitude star. NGC 4005 covers 45" by 30" of sky as seen in the 12.5-inch scope. It contains a faint nucleus and is elongated in p.a. 65°. **NGC 4000**, lying directly on the other side of the 8th-magnitude star, is a 15th-magnitude spiral that appears nearly edge-on. NGC 4000 measures 50" by 20" and is evenly illuminated across its surface.

Galaxies **NGC 3999** and **NGC 4011** lie closeby, just to the south. Each of these objects is an extremely faint, round smudge of light. A 16-inch telescope provides a glimpse of these gaalaxies, but don't expect to see much detail. Both galaxies have a photographic magnitude of 15.7 and measure less than 15" across.

A trio of easier galaxies lies 15' southeast. A probable elliptical, **NGC 4015** is the brightest of the three. Upon close inspection at high powers, NGC 4015 appears to have a faint, peculiar companion attached to its northern end. NGC 4015 has a bright nucleus and is reasonably large, measuring 50" by 35" in p.a. 75°. **NGC 4023** lies 8' southeast of NGC 4015. NGC 4023 is quite faint and covers 40" by 30" in size. **NGC 4021**, the trio's third component, is similar. It glows at photographic magnitude 15.3, spans 40" by 25", and is oriented almost exactly east-west.

The third grand galaxy group we'll look at this spring lies just over 1° northeast of the bright star 93 Leonis. The **NGC 3937 galaxy group** appears to be a small component of the Coma-Virgo Supercluster, along with such company as Abell 1367 and the Coma Cluster of galaxies. (Each of these magnificent clusters was described in *Deep Sky* 10, Spring 1985.) Seven members of the NGC 3937 group belong to the *NGC*, most of which are positioned along a curving string.

NGC 3937, the brightest object in the group, glows at photographic magnitude 14.0. NGC 3937 is visible in an 8-inch telescope on dark nights with a little effort, when it appears faint, small, and slightly elongated in p.a. 80°. The extremely faint and small galaxy **IC 2968** lies some 90" to the west.

NGC 3943 is located 5' northeast of a prominent 6th-magnitude star at the southern edge of the group. NGC 3943 is a faint, small, round galaxy only 25" in diameter. Fourteenth-magnitude **NGC 3947** can be spotted a little over 15' to the northeast. This 40"-by-35" spiral appears faint but contains a small, bright core. On extremely steady nights you may see a 16th-magnitude star at its eastern tip.

To the north is **NGC 3954**. This 14.4-magnitude elliptical galaxy is very small, faint, and exactly round. It is only 25" in size and has a bright nucleus. **NGC 3946** is 10' farther north. It is extremely faint and small even in a 17-inch telescope. Little detail can be detected of this 10"-diameter galaxy.

The bright elliptical galaxy **NGC 3940** is about 8' to the west. It shines at the equivalent brightness of a magnitude 14.3 star. It is 35" by 30" in size, contains a fairly bright core, and is oriented due east-west. **NGC 3929** precedes it by 10' to 12'. This 14th-magnitude elliptical galaxy appears very small and compact. It measures only 30" by 25" in size and contains a bright center. A magnitude-16 star is seemingly attached to its southern side.

The NGC 3937 group is a good hunting ground for anonymous galaxies. The following are telescopic descriptions of a few of the brighter ones using a 17.5-inch Newtonian telescope:

1. *#67*. Extremely faint and small, slightly elongated, 15" by 12" in p.a. 0°.
2. *#68*. Extremely faint, very small, slightly elongated, bright middle, diffuse, 35" by 30" in p.a. 90°.
3. *#71*. Extremely faint, extremely small, 15" by 12".
4. *#74*. Faint, very small, little elongated, pretty bright middle, 25" by 20" in p.a. 80°.
5. *#82*. Faint, small, almost round, gradually brighter to the middle, 25" by 20" in p.a. 100°. Two 16th-magnitude stars nearby.
6. *#73*. IC 742: Faint, pretty small, diffuse, faint middle, 45" diameter.
7. *#83*. Faint, small, slightly elongated, faint core, 25" by 20" in p.a. 110°.
8. *#72*. Very faint, small, little elongated, faint core, p.a. 90°, 25" by 20" in p.a. 90°. Between

Galaxy groups present new ground, unconquered territory for those who want to explore the limits of the galaxies they can observe.

two 14th-magnitude stars.
9. *#70*. Very faint, extremely small, pretty much elongated, bright middle, 20" by 10" in p.a. 125°. One 15th-magnitude star lies at the galaxy's northern tip.
10. *#92*. Faint, extremely small, round, bright core, 12" in diameter.
11. *#100*. Faint, small, slightly elongated, pretty bright middle, 30" by 25" in p.a. 50°.

Rood and Sastry would say that each of these galaxy groups is very different. The NGC 4169 group is small and very poor, curiously arranged in a rectangle. The NGC 4005 group contains a large number of edge-on spirals and is loosely spread out in no easily interpreted pattern. The NGC 3937 group strings along a curving line. Most of its members are ellipticals. What can you determine about any of these groups or galaxy clusters overall from your observations of three galaxy groups this spring?

Jeff Corder lives in Tice, Florida, and specializes in observing galaxies.

The M31 Globular Cluster System

Equipped with a large scope, you can spot twenty-one globular clusters suspended in the halo of the Andromeda Galaxy

by David Higgins

How often do you look at M31 during the course of an autumn observing season? Have you developed the attitude that once you've seen M31 there is just not much left to look at? Carefully observing and really *seeing* an object is different than merely looking at it. Soaking up all the information that an object, such as M31, has to offer requires patience, determination and practice. Of course knowing what you are looking at or searching for really helps. Having access to whatever material has been produced about a particular object and then observing when you get the chance goes a long way in providing you with the practice portion of observing deep-sky objects. Patience and determination you'll have to develop on your own.

M31 has intrigued and fascinated observers since the Persian astronomer Al-Sufi mentioned it in his *Book of the Fixed Stars*. Al-Sufi's A.D. 964 description of Andromeda's "Little Cloud," was the first recorded mention of the great nebula in Andromeda. Easily visible to the naked eye from a dark site, M31 was somehow overlooked by astronomers after Al-Sufi recorded it. Not until 1612 did the great nebula again gain the attention of an observer and find its way into a catalog describing nebulous objects. Using a small telescope, Simon Mayer, a German mathematician and astronomer, observed M31 and included it in his *Mundus Jovialis*, a pamphlet of descriptive observations.

Strangely, M31 again disappeared from the astronomical literature. Fifty years passed before it was included in a list of observations. In 1644 the French scientist Ishmael Boullian came upon M31 while looking for a comet. He included the great nebula in a publication that primarily discussed a long period variable star. Boullian was aware of the previous works of Al-Sufi and Mayer, and theorized that the nebula was variable due to its absence in the scientific publications.

Besides being one of the few real nebulous objects in the first deep-sky catalogs, M31 has the lone distinction of being the home of the first extragalactic supernova ever observed. One night in 1885 E. Hartwing used a 10-inch refractor at Dorpat observatory to look over M31 and observed a bright star only 16' from the core. Peaking at magnitude 5.4, nova S Andromedae was brighter than any normal nova that had been observed up to that time. At the time astronomers knew nothing of supernovae and to make matters worse, the nature and distance to "spiral nebulae" were unknown. Without this knowledge the distance and absolute magnitude of the nova could not be determined.

Although we take the spiral structure of M31 as an obvious fact, this was hardly the case in the 1880s. Spiral structure in nebulae had been observed and described by Lord Rosse as early as 1845. Using his 72-inch reflector Lord Rosse observed spiral structure in the face-on spiral M51. And even though M31 is much larger and closer it would take another forty years to detect spiral structure in Andromeda's great nebula. Because the galaxy is tilted only 12.5° from edge-on, observing spiral structure in M31 would prove almost impossible if not for a budding new technology.

Photography was about to explode in the 1880s. Photographing through a 20-inch reflector, a Welsh amateur astronomer named Isaac Roberts shot several plates in 1887 that for the first time recorded the whole disk of M31. Prior to this time the sensitivity of photographic plates was so low that exposure times had to be extremely long and were sensitive enough to record only the galaxy's bright nucleus. Not only did Roberts find the nucleus recorded on the plates but also dark lanes surrounding the core. Roberts also recorded numerous star images embedded in the galaxy's soft glow. The belief that spiral nebulae were solar systems in the making was so entrenched in astronomical thought that any understanding of the significance of Roberts' M31 plates would have to wait until Edwin Hubble advanced the course of astrophysics in the 1920s.

The debate over the distance and nature of spiral nebulae was going strong when Vesto M. Slipher made a discovery that would put astronomers on the road to solving this mystery. In 1912 M31 played a pivotal role in the discovery that the vast majority of spiral nebulae were receding from us at enormous speeds. M31 is approaching us at 300 kilometers per second, a speed that at the time of its discovery was the largest velocity ever observed. The fact that most of the spirals are receding from us at such speeds led some astronomers to suggest that these velocities imply large distances. Even though an increasing number of astronomers believed that Slipher's work

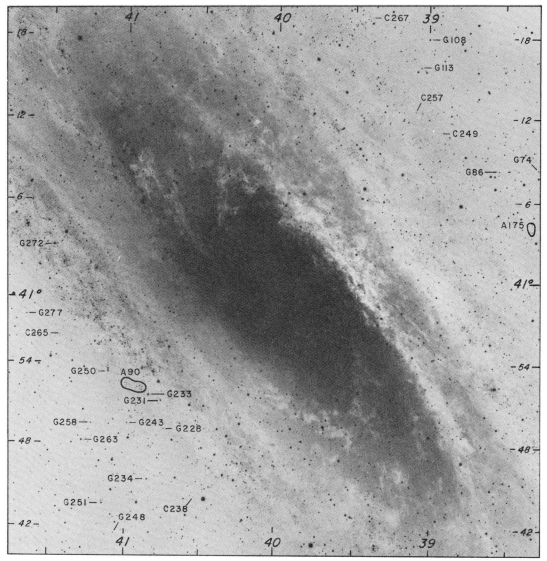

Chart 27 from Paul Hodge's Atlas of the Andromeda Galaxy shows a region around M31's core. Several clusters described in the article are visible here. Courtesy Paul Hodge.

went a long way in promoting the island universe theory, the distances to the spirals still were not known, and without this knowledge the relationship between the spirals and our own galaxy could not be determined.

Vesto M. Slipher also discovered that spirals are rotating. This was important evidence that would again lead astronomers to take seriously the island universe theory. But again the evidence was not conclusive enough to prove that the spirals were galaxies like our own Milky Way. The year 1912 was not only a year of discoveries that seemed to support the island universe hypothesis, it was also a year that saw the start of a study that became the dominating force behind proving that spiral nebulae were within and part of the Milky Way.

In 1912 Adrian van Maanan was appointed to the Mount Wilson staff were he undertook the project of measuring proper motions and parallaxes of stars. During his involvement in this project rotational and radial motions had been discovered in spiral nebulae. In 1915 van Maanan was asked to see if he could find any proper motions in the spiral M101. This project was along the same lines of his past studies so he undertook the project and proceeded to surprise himself and others in the process. Not only did van Maanen find proper motions in M101 but over the next ten years he found motions in a number of spirals. To detect these proper motions meant that spiral nebulae were close, so that many astronomers believed they were inside the Milky Way galaxy. Indeed many astronomers believed they were solar systems in the process of forming.

Although these data were accepted by most, studies conducted over the next ten years contradicted van Maanan's findings. But even with evidence mounting and questions that challenged the validity of van Maanan's data, several years passed before astronomers had hard evidence to support the island universe hypothesis. The final blow that indicated van Maanan's measurements were in error came when Edwin Hubble entered the scene. While photographing M31 and M33 in 1923, Hubble was able to resolve the outer spiral arms into individual stars. This was an exciting event to say the least but as a bonus he found a number of Cepheid variables, stars that had up to this time been found only in our galaxy and in the Magellanic Clouds. The light curves of Cepheids had been determined some years before and had been used by Harlow Shapley to determine the distances of the globular clusters which in turn gave us the outline of our own Milky Way galaxy.

By assuming that all Cepheids, regardless of where they are found, obey the same period-luminosity law, it was not difficult to determine the distance to M31. Although hesitant to publish his results at first Hubble knew that he had finally dealt a death blow to van Maanan's measurements of proper motions within the spirals. Published in 1924, Hubble's results caused quite an uproar. The majority of astronomers accepted and understood the

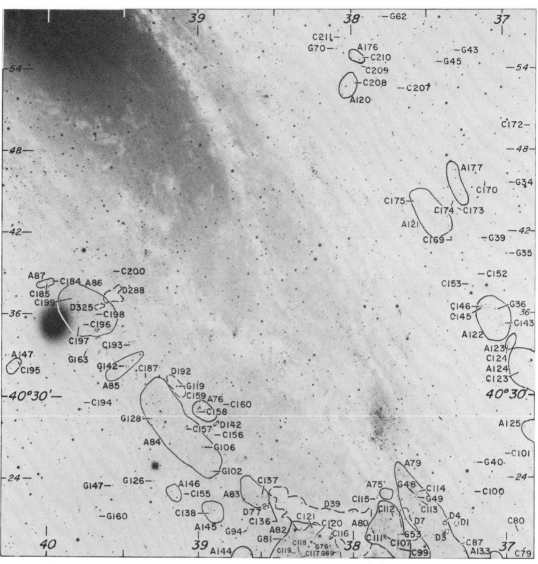

Chart 24 from the Hodge atlas shows the region of the M31 nucleus and satellite galaxy M32. Look for several of Higgins' clusters here.

significance of Hubble's findings but it took almost ten years for van Maanan's supporters to finally give way to the idea that the spiral nebulae were galaxies and the universe was home to countless numbers of Milky Ways.

Because M31 is so much like our own Milky Way it has its own compliment of deep-sky objects. Dating back to the time of Hubble, when it was determined that spiral nebulae were galaxies, studies have been made and continue to b e made, encompassing the many and varied objects that M31 contains. Of course these studies are published in the professional literature and difficult for most amateurs to find, and once found, hard to decipher. And even if you took a notion to try and find some of these objects finder charts for these faint objects are not always included so trying to figure out where and what to look for is also a problem. Astronomer Paul W. Hodge has alleviated this problem. His *Atlas of the Andromeda Galaxy* shows hundreds of objects of the same types in our own galaxy, much farther away and fainter of course.

The atlas consists of forty-one photographic charts that were taken at the prime focus of the 4-meter telescope of the National Observatory at Kitt Peak. The charts are divided into three sets that decrease in enlargement with two index charts that show the whole galaxy which in turn shows what area each one of main charts covers. The charts themselves are easy to use with a distinctive symbol that denotes each type of object plotted. A "g" is used to mark globular clusters, a "C" for open clusters, a solid outline for stellar associations, and a dashed outline for dust clouds. Plotted throughout the charts are 730 dust clouds, 188 stellar associations, 403 open clusters, and 355 globular clusters. Globular clusters have always fascinated me, so it didn't take long to put the atlas to good use. I spent the biggest part of the fall observing season searching for globulars with a 24-inch f/4 reflector and Hodge's atlas.

The dust clouds and stellar associations are very difficult if not impossible to observe. I think these objects are strictly in the realm of photography. The exception to that statement is the huge stellar association NGC 206. Easily visible in amateur scopes, NGC 206 lies at the southwestern tip of M31. The dust clouds are in the same category as the associations, very difficult if not impossible visually. There are open clusters visible in backyard scopes and I have found several but I have not devoted much time in tracking them down. That will be the subject for another project during next autumn's observing run. That leaves the brightest and easiest objects to observe in M31: the galaxy's extensive globular cluster system.

M31's globular cluster system is so much like our own that these extremely distant globulars give astronomers the opportunity to observe and test theories about the creation of the universe and the conditions that existed after the Big Bang. Globular clusters in M31 are metal poor, which means they contain a very low ratio of heavy

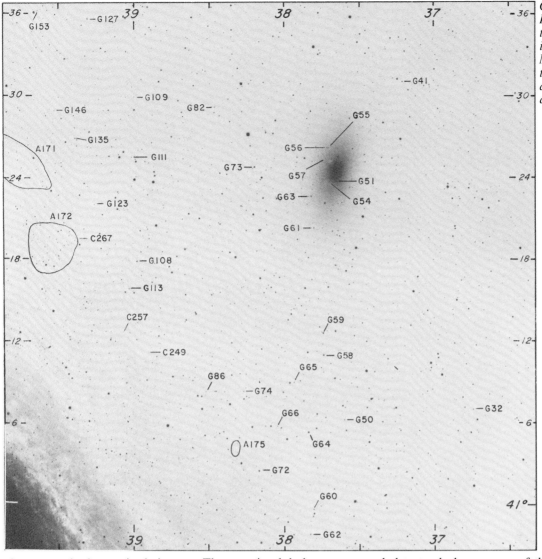

Chart 23 from the Hodge atlas shows the region surrounding satellite galaxy NGC 205. Several of the Higgins clusters are visible in this area.

elements to hydrogen in their stars. The stars in globulars are old Population II stars that have used up much of their hydrogen fuel. The globular clusters of M31, much like our own cluster system, orbit about the galactic center and occasionally each one plunges through the disk of M31. This sweeps out dust and gas, leaving it in the disk of M31 and stripping the globular clusters of heavy metals.

Visually the M31 globulars are no more that faint stars. Ranging from magnitude 13.7 to almost 19th magnitude, most of these globulars are not within the visual range of amateur scopes. But don't let the faint magnitudes dissuade you. Some of M31's globulars are visible in a 12.5-inch telescope and will provide you with a fascinating observing program. Whatever size telescope you are using, high magnification will be the key to spotting these clusters. You're not looking at a faint fuzzy, but rather a concentrated light source — what you're seeing is the central core of the globulars. Don't expect to see any of the surrounding stars around the periphery of the cluster. Remember that you will be looking through not only our galaxy's gas and dust but in some cases M31's.

These globulars are not plotted on any of the popular star charts and even if you had the coordinates most star charts have such small scales its impossible to plot them by hand. What you really need is Hodge's *Atlas of the Andromeda Galaxy*. This work will give you a good deal of the necessary information and large scale charts that are needed to track down some of these elusive clusters. Several other sources of information are available that will provide you with the rest of the information needed to round out the material required for successful observations. An article by Brian Skiff in the 1984 autumn issue of *Deep Sky* provides a great deal of information on this matter. Brian provides a number of objects in M31 and includes in his lists several globular clusters and their magnitudes. He also includes a photograph of M31 with different types of objects plotted on it. Although not as detailed as Hodge's plates, this photograph and the information included in the article form a good starting place until you can get your hands on the Hodge atlas.

An additional piece of material that contains a wealth of information is found in a paper published in the January 1985 issue of the *Astrophysical Journal*. Entitled "The M31 Globular Cluster System," and written by astronomers Crampton, Cowley, Schade, and Chayer, this paper contains much more information than the Hodge Atlas or Skiff's article. There are two major drawbacks to this paper, however. It is thick with technical jargon and contains no finder charts. Containing 509 objects identified as globular clusters, the paper does provide information on positions, magnitudes, and core radii. Also included is a general discussion about the globular cluster system and how they were identified. It also lists objects that in previous studies had been listed as globulars but have since been identified as starts. This paper is almost

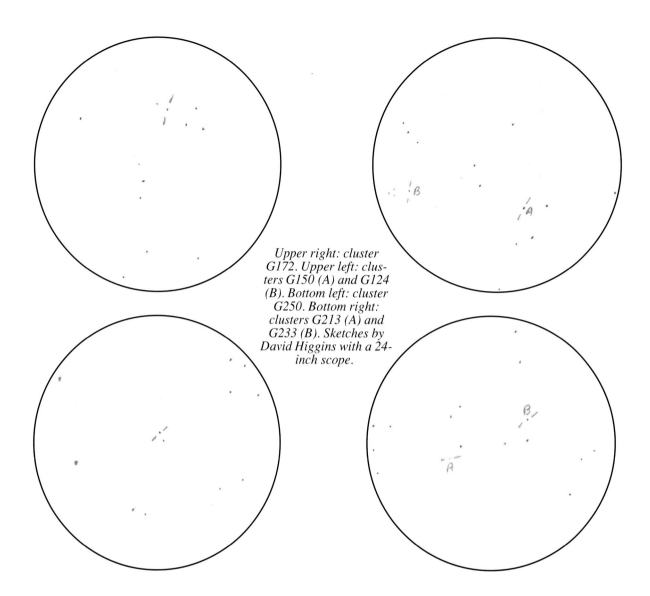

Upper right: cluster G172. Upper left: clusters G150 (A) and G124 (B). Bottom left: cluster G250. Bottom right: clusters G213 (A) and G233 (B). Sketches by David Higgins with a 24-inch scope.

a requirement for anyone interested in observing M31's globulars.

The descriptions of the following globulars are relatively simple and straightforward. What I've tried to do is give you an idea of how bright or faint these objects appear through a 24-inch f/4 scope. How they will appear to you depends on the size of telescope you are using, the observing conditions, and your observing site. Also included in the descriptions are the cluster magnitudes and core diameters in arcseconds. You must keep several things in mind as you track these faint objects down. One item in particular is that the globulars along the periphery of the galaxy are a little easier to see than those nearer the center. Clusters in or near the core are very reddened by dust and gas.

It has often been said that regardless of the magnitudes quoted or information given on deep-sky objects, it is always a good idea to be a little skeptical of this data. There are so many variables in observing and published descriptions of deep-sky objects that the best advice for observers is simply do your own observing.

The M31 Globular Cluster System

G35. Magnitude 15.6, size 2.9".
Very faint with averted vision required. Fades in and out with the seeing. 9mm Nagler at 270x.

G87. Mag. 15.6, size 2.9".
Very faint with averted vision required for best view. 9mm Nagler at 270x.

G78. Mag. 14.3, size 3.2".
Bright and easily seen. Held steady with direct vision. This is the brightest globular of the group. In the same eyepiece failed with G70. 9mm Nagler at 270x.

G70. Mag. 16, size 2.3".
Extremely faint and difficult. Fades in and out with averted vision required. Same eyepice field with G78. 7mm Nagler at 348x.

G96. Mag. 15.5, size 2.7".
Even with a magnitude of 15.5, G96 was easily seen. Averted vision gave the best view. 9mm Nagler at 270x.

G119. Mag. 15, size 2.7".
Faint but held steady with direct vision. In the same eyepiece field with M31's companion galaxy M32. 9mm Nagler at 270x.

G134. Mag. 15.9, size 2.7".
Faint with averted vision required for best views. Could be seen with direct vision when conditions permitted. 13mm Nagler at 187x.

G156. Mag. 15.6, size 2.5".
Faint with averted vision required. Held steady with direct vision when conditions permitted. In same eyepiece field with M32 and G176. 9mm Nagler at 270x.

G176. Mag. 16.3, size 2.7".
Extremely faint with averted vision required to even catch a glimpse of it. Bright star to the west has a tendency to wash it out. In same eyepiece field with M32 and G156. 7mm Nagler at 348x.

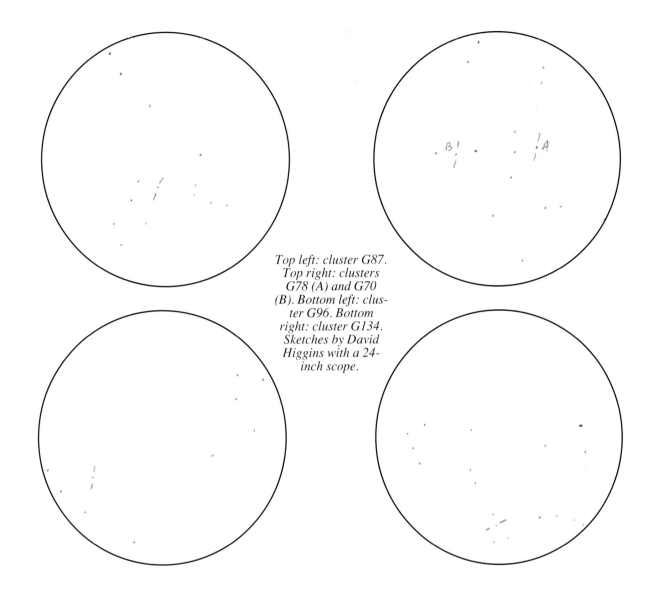

Top left: cluster G87. Top right: clusters G78 (A) and G70 (B). Bottom left: cluster G96. Bottom right: cluster G134. Sketches by David Higgins with a 24-inch scope.

G172. Mag. 15.2, size 2.4".
Very faint with averted vision required. Fades in and out with the seeing conditions. 9mm Nagler at 270x.

G150. Mag. 15.4, size 2.7".
Very faint with averted vision required. In same eyepiece field with G124. 9mm Nagler at 270x.

G124. Mag. 15.7, size 2.3".
Very faint and difficult with averted vision required. Fades in and out with seeing conditions. 9mm Nagler at 270x.

G250. Mag. 16, size 3.4".
Extremely faint with averted vision required to catch a glimpse of it. 13mm Nagler at 187x.

G213. Mag. 14.7, size 2.5".
Easily seen with direct vision. In the same eyepiece field with G233. 13mm Nagler at 187x.

G233. Mag. 15.4, size 2.6".
Faint and difficult. Averted vision required. In the same eyepiece field with G213. 13mm Nagler at 187x.

G234. Mag. 15.9, size 2.7".
Faint with averted vision required to see well. In the same eyepiece field with G263. 13mm Nagler at 187x.

G263. Mag. 15.5, size 2.5".
Faint with averted vision required to see well. In the same eyepiece field with G234. 13mm Nagler at 187x.

G244. Mag. 15.4, size 2.6".
Faint with averted vision required for best view. Fades in and out with the seeing conditions. In the same eyepiece field with G226. 9mm Nagler at 270x.

G226. Mag. 15.5, size 3.8".
The same as G244, faint with averted vision required. It also fades in and out with the changing conditions. 9mm Nagler at 270x.

G279. Mag. 15.4, size 4.9".
Although the largest in size of all of the globulars that I've observed, this cluster was very faint. Averted vision was required, and it suffers from fading in and out. 13mm Nagler at 187x.

G73. Mag. 15, size unknown.
I've included G73 although it is not associated with M31. It is a globular cluster in M31's companion galaxy NGC 205. 9mm Nagler at 270x.

Regardless of how many clusters you find or don't find keep in mind what you are seeing. Besides observing these clusters, spend some time studying the globular cluster system of our own galaxy and how it interacts with the Milky Way. By trying to understand the globulars in our own galaxy it will provide you with a much better appreciation of what you are seeing when you observe the clusters in M31.

David Higgins is the central coordinator of the National Deep Sky Observers Society, a contributing editor of this magazine, and frequently observes with large telescopes near his home in Krebs, Oklahoma.

Deep Sky Reports
The Galaxies of Orion
by Steve Gottlieb

Now wipe that silly grin off your face because I'm quite serious about this topic. Say the word "galaxies" and it's unlikely that Orion is going to come to mind. Most observers rightly think of Orion in terms of its beautiful emission, reflection, and dark nebulae, but it is also home to quite a number of faint galaxies and even some unexpected treats. So follow along and you'll discover there is much more to the Great Hunter than just M42 and the Horsehead Nebula. You won't find any of these objects plotted on Tirion's *Sky Atlas 2000.0* because they fall just below the magnitude cutoff. Instead, you'll need to use the more detailed *Uranometria 2000.0* as your guide. Most of my observations were made with a 13.1-inch Odyssey 1 at 144x and I later returned to pick up some of the fainter galaxies with my 17.5-inch f/4.4 Dobsonian operating at 220x. The best place to start hunting for galaxies is to the west of the galactic plane in the southwest portion of the constellation where Orion blends into the rich galaxy fields of Eridanus.

The barred spiral **NGC 1691** can be easily found 50' north of 8 Orionis (also known as Pi^5 Orionis), a 4th-magnitude star. At 166x, it appeared faint and very small with either a bright stellar core (magnitude 12 to 13) or else a foreground star is superposed. Backtracking to 8 Orionis and moving 50' south brings you to a much fainter elliptical galaxy, **NGC 1690**. Dominating the field is a magnitude 6.6 star 7' to the southwest and several faint stars surround the galaxy, including one at the western edge. At B magnitude 15, this object was very dim in my 13-inch but held steadily with averted vision and appeared as a very small, round, featureless disk.

Just over 2° southeast you'll find a trio of faint galaxies. The brightest is **NGC 1713**, a fairly faint and small elliptical with a gradually brighter center surrounded by a faint halo. In 1854 Lord Rosse discovered a very faint companion located 2.7' west-northwest that was missed by both John and William Herschel. As **NGC 1709** is both dim and diminutive, it's easy to see why, but a 13th-magnitude star just 45" west can help pinpoint its location. Moving 20' northeast will bring you to **NGC 1719**. This faint spiral is also very small yet clearly elongated east-west. Look for a faint star, magnitude 14 or 15, at the west end.

In the southwest corner of Orion, you'll find a group of six NGC galaxies located less than 2° east of Mu Eridani. **NGC 1670** was easy to see and hold steady with direct vision. Although small and round, it contains a brighter core and a magnitude 14 star is close southeast. Moving 28' east-northeast you'll come to **NGC 1678**, a slightly fainter oval galaxy with a small bright core. A brighter magnitude 12 star sits prominently just 40" to the west.

Travel 30' southeast and you'll arrive at a clump of four faint galaxies, **NGC 1682**, **NGC 1683**, **NGC 1684**, and **NGC 1685**. The brightest in the group, NGC 1684, appears as a fairly prominent oval oriented east-west with a large bright core. A magnitude 9 star 3' south highlights the field. Second brightest is the close neighbor NGC 1682, 3' to the west. Although fairly faint, it had a bright core and a stellar nucleus. The faintest member of the group, NGC 1683, is 5' further north. This dim, low surface brightness galaxy was found hiding among an elongated group of five magnitude 14 stars. Finally, NGC 1685 is also located in the same field 5' to the northeast. Look for a slight elongation northwest-southeast and a magnitude 14 star off the southeast edge.

While you're here near the Eridanus border, let's locate **NGC 1661** to the northwest. This faint, small spiral appears roundish and contains a brighter core. Look for a line of four stars oriented east-west situated just 2' south. As a side trip, you can cross the Eridanus border just 20' to the west and take a peek at the pair of galaxies, **NGC 1654** and **NGC 1657**.

Now head back to the NGC 1684 group and then turn your scope about 2° east-southeast. You'll come to a loosely knit trio whose brightest member is **NGC 1729**. In my 17.5-inch at 220x, it appeared elongated roughly north-south and was bracketed by an 11th-magnitude star off the east end and a 13th magnitude star off the northwest side. About 25' east you'll find a similar galaxy, NGC 1740, which is elongated northeast-southwest and contains a small bright core. At the southwest edge is positioned a star between 12th and 13th magnitude. Finally, look just 10' east and you may glimpse the very dim **NGC 1753**. This galaxy required averted vision to see well and a close, faint double star, perhaps composed of magnitude 13 and 15 components, was visible 3' southeast.

If you now return to 8 Orionis, where we started, and then move 80' southeast, you'll come to the 4.5 magnitude star 10 Orionis. Another hop 50' east brings you to **ADS 3623**, a pleasing magnitude 6.5 and 7.7 duo with a separation of 14". **NGC 1762** can then easily be found by drifting 1.6' from this pair (or moving 24' due east). The drifting method works perfectly on altazimuth scopes and floating through my field was a faint, small oval galaxy, extended north-south with a magnitude 13 or 14 star attached at the east end. Another spiral, **NGC 1819**, is located 43' west of a magnitude 5.5 star about 4° northeast of NGC 1762. This small galaxy was elongated southeast-northwest (position angle 120°) and contained a noticeably brighter core.

Of course, no trip to Orion would be complete without a deep, long gaze at the breathtaking Orion Nebula, M42. Take your time but when you're done move your scope just 2° west for a big surprise — **NGC 1924**. I'm amazed that a fairly bright galaxy this close to M42 has not been mentioned often before. NGC 1924 appeared moderately bright and large (the best of the Orion galaxies) with a slightly brighter core. It was bracketed by a magnitude 9 star 4' northwest and a magnitude 8 star 4' east. Next time you're gazing into the sword of Orion, take a look at this neglected galaxy.

The northeast portion of Orion flows into the dusty Milky Way where galaxies are generally blotted out from view but 4.5° north of Betelgeuse, you'll run across the solitary galaxy **NGC 2119**. As expected, the star fields are rich here but the galaxy was not too difficult to pick out in my 17.5-inch scope at 220x. A small oval oriented northwest-southeast with a bright core was visible and a 10th-magnitude field star was positioned 2' northeast.

For the last leg of our journey, head straight to Bellatrix (Gamma Orionis), the upper right star in the outline of the hunter. Just under 1° west-northwest you may pick up a very small, round object catalogued as **NGC 1875**. A faint stellar nucleus was visible and a 14th-magnitude star could be seen about 1' west. I later noticed on the *POSS* that an interacting triplet of galaxies is located just southeast. Anyone want to give these a try?

We've now run across several small groups. Just over 1° west of NGC 1875 you'll run into a very different animal, a rich cluster of galaxies. First catalogued in 1957 by George Abell as number 539 in his list of 2712 galaxy clusters, its brightest members shine at only 15th magnitude (photographic). But this cluster is worth the effort because eight members were located in my 17.5-inch in one small field. None of these galaxies are listed in either the *NGC* or *IC* and so you'll need to refer to Zwicky's *Catalogue of Galaxies and Clusters of Galaxies* (*CGCG*) for identifications. The brightest member, **Z421-18**, is actually a triple system and is part of VV 161, a chain of five or six galaxies. With close scrutiny, I could resolve two objects, oriented north-south. In addition, several other nearby members of this chain were visible with averted vision. Search this field carefully, as these objects are quite small and dim and could be passed over for extremely faint stars even at moderate power.

The next clear night you're observing, why not take a break from the usual fare of open clusters and nebulae along the winter Milky Way and head out into Orion for some galaxy hunting? I guarantee it will spice up your evening and if the transparency is good you should find success with most of these galaxies in a 10-inch or 12-inch scope.

Nearly all the information in the table is from the *Third Reference Catalogue of Bright Galaxies* (*RC3*) by Gérard de Vaucouleurs, Antoinette de Vaucouleurs, Harold G. Corwin, Jr., Ronald J. Buta, Georges Paturel, and Pascal Fouqué (Springer-Verlag, New York, 1991). A few approximate sizes and magnitudes are from the *UGC*, *CGCG*, and *MCG*. The magnitudes are total B magnitudes; as you can see for the three objects with V magnitudes given in the last column, the visual magnitudes will typically be a magnitude brighter. Thee are still moderately faint, but all the objects have high surface brightnesses, making them relatively easy to pick up at high power. □

Steve Gottlieb is an enthusiastic galaxy observer living in Albany, California. He is a frequent contributor to Deep Sky.

Galaxies in Orion

Designation	R.A.(2000.0)Dec.		Type	Size	Mag.	Sfc.	P.A.	Chart
NGC 1661	4h47.1m	-2°03'	SA(s)bc:II-III	1.4' by 0.9'	14.0	14.1	35°	224
NGC 1670	4h49.7m	-2°46'	SA0⁰	2.1' by 1.0'	13.7	14.3	112°	224
NGC 1678	4h51.6m	-2°37'	SA0⁰? pec	1.1' by 0.8'	14.2	13.8	110°	224
NGC 1683	4h52.3m	-3°01'	—	—	—	—	—	224
NGC 1682	4h52.3m	-3°06'	—	0.3'	14p	—	—	224
NGC 1684	4h52.5m	-3°06'	E+: pec	2.5' by 1.7'	13p	—	—	224
NGC 1685	4h52.6m	-2°57'	SB(r)0/a	1.3' by 0.9'	14.5p	—	—	224
NGC 1690	4h54.3m	+1°38'	E:	1.0' by 0.3'	14.8	13.5	116°	224
NGC 1691	4h54.6m	+3°16'	(R)SB(s)0/a:	1.7' by 1.5'	13.0	13.9	85°	224
NGC 1709	4h58.7m	-0°29'	SB0⁰:	0.9' by 0.7'	15.2	14.6	10°	224
NGC 1713	4h58.9m	-0°29'	E+:	1.4' by 1.2'	13.7	14.3	—	224 (V=12.6)
NGC 1719	4h59.6m	-0°16'	Sa: sp	1.1' by 0.3'	14.5	13.2	102°	224
NGC 1729	5h00.3m	-3°21'	SA(s)c I-II	1.6' by 1.3'	13p	—	150°	224
NGC 1740	5h01.9m	-3°18'	S0-:	1.5' by 1.2'	15p	—	125°	224
NGC 1753	5h02.5m	-3°21'	(R')SBa? pec	1.4' by 0.6'	15.5p	—	15°	224
NGC 1762	5h03.6m	+1°34'	SA(s)c: II	1.7' by 1.1'	13.3	13.9	175°	224 (V=12.6)
NGC 1819	5h11.8m	+5°12'	SB0	1.7' by 1.2'	13.4	14.0	120°	225
NGC 1843	5h14.1m	-10°38'	SAB(s)cd: II-III	1.6' by 1.2'	13.3	13.8	50°	270
Z421-15	5h16.3m	+6°26'	—	—	15.5z	—	—	180
Z421-16	5h16.5m	+6°23'	—	—	15.6z	—	—	180
Z421-17	5h16.6m	+6°30'	—	—	15.6z	—	—	180
Anon	5h16.6m	+6°29'	—	—	16:p	—	—	180
Z421-18a	5h16.7m	+6°26'	—	—	15.1z	—	—	180
Z421-18b	5h16.7m	+6°26'	—	—	15.1z	—	—	180
Anon	5h16.8m	+6°29'	—	—	16:p	—	—	180
Z421-19	5h17.0m	+6°33'	—	—	15.6z	—	—	180
NGC 1875	5h21.8m	+6°41'	SaO-:	1.6' by 0.4'	14.6	14.1	—	180
NGC 1924	5h27.9m	-5°19'	SB(r)bc II-III	1.6' by 1.2'	13.3	13.8	50°	225 (V=12.5)
NGC 2110	5h52.2m	-7°27'	SaO-	1.7' by 1.3'	—	—	20°	271
NGC 2119	5h57.4m	+11°57'	E	(1.2' by 1.0')	15:p	—	—	181

Did you know about the bright galaxy NGC 1924, an object located just 2° west of the Orion Nebula? Jack Marling's photo shows the galaxy's position (arrow). Exposure data: 200mm f/2.8 lens, hypersensitized Tech Pan film, hydrogen-alpha filter, 40-minute exposure.

The Coma Bernices and Abell 1367 Galaxy Clusters

by Brian Skiff

Want to find a lot of deep-sky objects without a lot of searching? Try looking for nearby groups and clusters of galaxies. These range from incredibly dense clouds of very faint fuzzies down to groups of only a handful of spirals and ultimately simple pairs of galaxies orbiting about a common center, or perhaps colliding.

That galaxies cluster was recognized by the turn of this century, even before the "spiral nebulae" were known to be outside the Milky Way. Charles Messier himself found three dozen galaxies in what is now called the Virgo cluster. Photographers began turning up the brightest rich galaxy clusters around 1900. In 1906 astronomer Max Wolf in Heidelberg, Germany "discovered" the Coma Berenices galaxy cluster on plates he took with a 16-inch astrograph. Though several dozen galaxies had already been catalogued there mostly by visual observers, Wolf's photographs showed for the first time many hundreds of cluster members.

Before World War II, there was evidence not only that perhaps all galaxies belonged to groups or clusters, but also that the clusters themselves were lumped together. Harlow Shapley at Harvard found this "metagalactic structure" — as he termed it — on southern hemisphere wide-field survey plates. Edwin Hubble, working in the north at Mount Wilson, counted galaxies in about 1300 tiny fields photographed with the 60- and 100-inch telescopes. He concluded that this could not be explained by a random clumping of isolated "field" galaxies, but that galaxies must occur predominantly in groups and clusters.

Galaxy-counting continued into the 1960s, when the Lick Observatory survey was completed by Donald Shane and Carl Wirtanen. They assayed about a *million* galaxies to beyond magnitude 18 in the northern two-thirds of the sky, and called attention to prominent "clouds" of galaxies and much larger superclusters. Using the National Geographic-Palomar Observatory *Sky Survey* (POSS), George Abell of UCLA made a catalogue of 2700 rich clusters of galaxies. Palomar's Fritz Zwicky catalogued about 10,000 clusters of all kinds north of the celestial equator. Taken together, these surveys conclusively showed that galaxy clusters are far more numerous than almost anyone had believed.

The Abell catalogue is now the only widely used list because it includes the easily studied populous objects. He included clusters which have more than 50 members no more than two magnitudes fainter than the third-brightest cluster galaxy. Abell also restricted membership to those galaxies within an angular diameter that varied with the magnitude of the brightest cluster galaxies. This second criterion kept the circle containing the cluster to a fixed size in space, approximately 6 megaparsecs.

Though sophisticated classification schemes exist, galaxy clusters can be separated into three broad categories: "regular" or "compact," "intermediate," and "irregular" or "open."

The photographs you see in magazines of tremendous aggregations of galaxies all portray regular clusters. These symmetrical objects are usually nearly circular in outline, and are often dominated by one or a few supergiant elliptical galaxies located at the center. The component galaxies include very few spirals. The well-known Coma Berenices and Corona Borealis clusters fall into this class.

The open galaxy clusters are unconcentrated and show an irregular outline. Here spirals are as numerous as elliptical and lenticular types. The nearby Virgo cluster and the Hercules galaxy cluster (Abell 2151) are typical. The various branches of the Virgo cluster extending into Leo, Canes Venatici, and elsewhere are typical of irregular clusters, where several nuclei or "subclusters" are present.

The intermediate clusters are just that: the constituent galaxies are somewhat concentrated, but generally not in a spherical form — lenticular galaxies are most numerous. The Perseus cluster (Abell 426), whose brightest galaxies are arranged in a string, is an example.

The true size, shape, and population of any cluster of galaxies are difficult to determine. If most galaxies are in clusters or groups, how can we tell the boundary between a cluster and the "field"? Do galaxies simply thin out and merge into the adjacent group? Look at the Coma cluster: the literature says it's 12° across; Thomas Noonan reanalyzed Zwicky's own data and found a diameter of only 200 arcminutes (3.3°); and Abell showed most of the Coma galaxies lie in a 5.4°x4.8° oval! These seeming discrepancies arise from how the "background" of galaxies was determined. Even using redshifts to pick out members is not conclusive. The range of velocities is large even for near-certain members due to the motions of individual galaxies. In the Coma cluster, the average redshift is about 7000 km/sec, but galaxies with velocities ranging from about 5000 to 9000 km/sec belong to the group.

As more and more velocities were measured in the Coma cluster, Stephen Gregory and Laird Thompson found in 1978 that a band of galaxies at the same redshift arcs westward from the Coma group to another rich, but irregular cluster (Abell 1367) about 20° away in Leo. The two clusters lie near opposite ends of a supercluster stretching perhaps 100 megaparsecs. No wonder the edge of the Coma cluster is hard to pin down!

These two clusters make a nice pair for visual observing, as they contain some of the richest fields of galaxies in the sky. The Coma cluster, also known as Abell 1656, is the classic rich, well-concentrated cluster containing two supergiant elliptical galaxies at its center and a few spirals on the periphery of the densest region. The central density of galaxies is about ten times higher than that for areas a few degrees away. Several of the individual galaxies are radio sources, and there is a general enhancement of the radio background, which is evidence for a gaseous intracluster medium. The cluster is also an X-ray source.

Less spectacular, Abell 1367 is an irregular cluster containing a mix of spiral and elliptical galaxies scattered in small clumps. It is a diffuse source of radio radiation; several discrete X-ray sources have been located within the cluster. Two of the latter were recently shown to be quasars near the brightest central galaxy, NGC 3842 (see *Deep Sky 9*, page 38). The two quasars are in the data list for the cluster as GP 945 and GP 872; though not accessible visually (they're magnitude 19), it shouldn't be difficult for amateurs to photograph them with modest equipment.

My visual observations of these objects in February and June 1984 were made with the U.S. Naval Observatory's 24-inch Cassegrain telescope stopped down to an equivalent 12½-inch aperture. A 24mm Konig eyepiece gave about 225x. Its finder, a 5-inch refractor, served also for views of the brighter objects. Additional observations were made with 6- and 10-inch Newtonians.

The Coma Cluster

The brightest galaxies in the Coma cluster are **NGC 4889**, just east of the center of the cluster, and **NGC 4874**, slightly fainter, 9.2′ away to the west. Both are visible in the 5-inch refractor at medium power. NGC 4874 is more difficult to view, appearing as a poorly-concentrated roughly circular patch. A magnitude 12.5 star lies 2′ southwest. A magnitude 7 star only 6.5′ north of the galaxy interferes somewhat with viewing here and in larger apertures. With the 12½-inch, NGC 4874 is about 1.5′ across and roughly circular, but many tiny galaxies nearby make for a seemingly irregular outline. The outer regions have a moderate even concentration; the core arises in a slight, sharp condensation 20″ across, but there is no distinct nucleus.

NGC 4872 is the closest in a virtual cloud of tiny objects around NGC 4874. Lying at about 50″ in p.a. 200° from NGC 4874, the 12½-inch shows a little star-like point with a weak halo no more than 15″ in diameter. **NGC 4871**, 1.3′ west of NGC 4874, is a small concentrated spot similar to NGC 4872, but somewhat fainter. NGC 4873, at 1.6′ in p.a. 340°, seems larger and less concentrated than the others. The well-defined portion is about 15″ across, elongated a bit east-southeast/west-northwest. IC 3998, 2.3′ away in p.a.

Galaxies in Abell 1656
12h59m43s, +27°58′14″ (2000)

Type	GMP	GP	NGC/IC	RA (2000)	Dec	Mag.	SB	Size
E	4943	888	—	12h57m20s	+27°29′42″	14.5	—	0.3′ x 0.2′
SA0⁻	4928	886	NGC 4839	12h57m22s	+27°30′00″	12.4	14.6	4.2′ x 2.1′
E1:	4829	856	NGC 4840	12h57m30s	+27°36′42″	13.5	12.2	0.6′ x 0.6′
SA0 pec	4822	857	NGC 4841A	12h57m30s	+28°28′42″	12.7:	13.3	1.7′ x 1.7′
E0 pec	4806	854	NGC 4841B	12h57m32s	+28°29′00″	13.3:	14.0	1.3′ x 1.3′
E0	4749	850	NGC 4842A	12h57m33s	+27°29′42″	13.9	12.4	0.6′ x 0.4′
E3:	4792	849	NGC 4842B	12h57m33s	+27°29′12″	15.0	12.7	0.4′ x 0.3′
Sa: sp	4471	763	NGC 4848	12h58m04s	+28°14′30″	13.6	13.5	1.8′ x 0.6′
SA0	4315	726	NGC 4850	12h58m19s	+27°58′07″	14.1	14.9	1.6′ x 1.6′
E	4313	727	NGC 4851	12h58m19s	+28°09′01″	14.6	—	0.6′
E	4308	722	IC 839	12h58m20s	+28°09′13″	15.3	—	0.5′
E	4230	710	—	12h58m28s	+28°00′55″	13.8	—	1.5′ x 1.5′
(R′)SA0⁻?	4156	690	NGC 4853	12h58m33s	+27°35′49″	13.1	13.2	1.2′ x 1.0′
Sa?	4130	—	IC 3943	12h58m34s	+28°06′49″	15.6ᴮ	—	0.7′
SB0	4017	660	NGC 4854	12h58m45s	+27°40′31″	13.9	—	1.2′
SA0 sp	3997	655	IC 3946	12h58m46s	+27°48′37″	13.9	12.9	1.0′ x 0.5′
SA0	3958	643	IC 3947	12h58m50s	+27°47′07″	14.6	—	0.6′
SA0 pec sp	3896	627	IC 3949	12h58m54s	+27°50′01″	14.1	12.6	1.3′ x 0.2′
S0?	3818	613	—	12h58m60s	+28°13′31″	14.1	—	0.7′
SBb	3816	612	NGC 4858	12h58m60s	+28°06′55″	15.0	12.4	0.4′ x 0.3′
E2:	3792	—	NGC 4860	12h59m02s	+28°07′25″	13.5	13.2	1.0′ x 0.9′
SBa	3779	607	—	12h59m03s	+27°38′43″	14.7	13.4	0.6′ x 0.5′
SA0	3761	602	IC 3955	12h59m04s	+27°59′49″	14.3	—	0.9′
E	3739	595	IC 3957	12h59m04s	+27°46′07″	14.5	—	0.4′ x 0.4′
SB0	3733	596	IC 3960	12h59m06s	+27°51′19″	14.5	12.4	0.4′ x 0.4′
E3	3730	594	IC 3959	12h59m06s	+27°47′02″	13.9	12.7	0.6′ x 0.6′
E2	3664	583	NGC 4864	12h59m11s	+27°58′38″	13.6	12.1	0.6′ x 0.4′
SA0 sp	3660	580	IC 3963	12h59m11s	+27°46′32″	13.5	12.7	0.6′ x 0.3′
S0	3656	575	—	12h59m12s	+28°04′38″	14.4	12.5	0.4′ x 0.4′
E3	3639	575	NGC 4867	12h59m13s	+27°58′14″	14.2	14.2	1.1′ x 1.1′
E6	3561	560	NGC 4865	12h59m18s	+28°05′02″	13.3	13.4	1.4′ x 0.8′
E6	3510	544	NGC 4869	12h59m21s	+27°54′38″	13.5	13.7	1.1′ x 1.1′
dbl galaxy	34I84	534	—	12h59m26s	+28°58′26″	—	—	0.6′
E6	3423	514	IC 3976	12h59m27s	+27°51′02″	14.4	—	0.7′
SA0	3414	511	NGC 4871	12h59m28s	+27°57′20″	13.9	12.1	0.5′ x 0.4′
SB0	3400	502	IC 3973	12h59m29s	+27°53′02″	14.2	—	0.9′
SA0	3367	499	NGC 4873	12h59m30s	+27°59′02″	14.7	13.1	0.8′ x 0.6′
SB0	3352	496	NGC 4872	12h59m32s	+27°56′50″	13.7	13.7	1.0′ x 1.0′
E⁺0	3329	489	NGC 4874	12h59m34s	+27°57′32″	11.9	14.0	2.7′ x 2.7′
SA0	3296	477	NGC 4875	12h59m36s	+27°54′26″	14.7	—	0.7′
E5	3201	452	NGC 4876	12h59m42s	+27°54′44″	14.2	12.2	0.4′ x 0.4′
SB0	3170	442	IC 3998	12h59m45s	+27°58′26″	14.7	—	0.6′ x 0.4:′
Sa?	3165	441	—	12h59m45s	+27°42′38″	13.9	—	1.2′ x 0.3′
SB0	3073	423	NGC 4883	12h59m54s	+28°02′02″	14.3	—	0.4′ x 0.3:′
SA0⁻	3055	416	NGC 4881	12h59m56s	+28°14′50″	13.5	13.5	1.0′ x 1.0′
E0	2975	385	NGC 4886	13h00m02s	+27°59′15″	13.9	13.5	0.8′ x 0.8′
S0	2946	376	—	13h00m04s	+27°58′45″	14.9	—	0.9′
E0	2940	368	IC 4011	13h00m04s	+28°00′15″	15.0	—	0.4′ x 0.4:′
E3	2922	360	IC 4012	13h00m06s	+28°04′45″	14.8	—	0.4′ x 0.3:′
E⁺4	2921	362	NGC 4889	13h00m06s	+27°58′39″	11.4	13.4	3.0′ x 2.1′
E0	2839	332	IC 4021	13h00m13s	+28°02′33″	14.9	12.2	0.3′ x 0.3′
SA0	2815	329	NGC 4894A	13h00m14s	+27°58′03″	13.9	—	0.8′
E pec	2798	322	NGC 4898A	13h00m15s	+27°57′21″	13.6	—	0.9′
SA0 pec sp	2795	326	NGC 4895	13h00m16s	+28°12′09″	12.8	13.4	2.3′ x 0.9′
E pec	2794	319	NGC 4898B	13h00m16s	+28°12′09″	15.3	—	0.6′
SB0	2727	297	IC 4026	13h00m16s	+27°57′27″	14.7	—	0.5′ x 0.4:′
S0⁻ pec	2629	—	NGC 4896	13h00m20s	+28°02′51″	13.7	13.4	1.3′ x 0.7′
Sdm:	2559	251	IC 4040	13h00m33s	+28°20′45″	14.7	14.1	1.1′ x 0.6′
E3	2541	245	NGC 4906	13h00m36s	+28°03′33″	14.2	—	0.6′ x 0.5:′
E6	2535	241	IC 4041	13h00m38s	+27°55′27″	14.5	13.7	0.7′ x 0.7′
SB0	2516	237	IC 4042	13h00m39s	+27°59′51″	14.3	—	0.6′ x 0.5:′
SB(r)b	2441	215	NGC 4907	13h00m41s	+27°58′21″	13.4	13.9	1.5′ x 1.3′
E4	2440	217	IC 4045	13h00m47s	+28°05′27″	13.9	12.6	0.7′ x 0.5′
E5	2417	208	NGC 4908	13h00m50s	+28°02′33″	13.5	13.7	1.1′ x 1.0′
S0	2413	209	—	13h00m49s	+28°21′57″	13.8	—	0.8′
Sm?	2393	198	—	13h00m52s	+27°47′03″	15.3	—	0.6′ x 0.4′
E0	2390	195	IC 4051	13h00m52s	+28°00′27″	13.4	13.8	1.3′ x 1.1′
SAB(r)bc	2374	192	NGC 4911	13h00m54s	+27°47′27″	12.8	13.1	1.3′ x 1.2′
E	2252	167	—	13h01m07s	+27°49′10″	14.8	—	0.8′
(R⁻)SA(r)0?	2157	149	NGC 4919	13h01m16s	+27°48′34″	13.7	13.7	1.4′ x 0.8′
SB(rs)ab	2059	132	NGC 4921	13h01m23s	+27°53′10″	12.1	14.0	2.7′ x 2.4′
(R⁻)SA(r)0⁻?	2000	120	NGC 4923	13h01m29s	+27°50′52″	13.6	13.4	1.3′ x 1.2′
S0 sp	1853	94	—	13h01m44s	+28°05′46″	14.2	—	0.6′
SA0⁻	1750	64	NGC 4926	13h01m52s	+27°37′35″	12.9	13.4	1.4′ x 1.3′
SA0⁻	1715	58	NGC 4927	13h01m55s	+28°00′29″	13.6	—	1.2′
S0 pec?	1616	44	NGC 4926A	13h02m05s	+27°38′59″	14.4	13.5	0.8′ x 0.6′

Column 1: revised Hubble galaxy types from the *Second Reference Catalogue of Bright Galaxies*, by G. and A. de Vaucouleurs and H. Corwin, Univ. of Texas Press, 1976. **Column 2:** catalogue numbers from the J. Godwin, N. Metcalfe, and J. Peach galaxy listing in *Monthly Notices of the R.A.S.*, 202:113+microfiche (1983). **Column 3:** catalogue numbers from the J. Godwin and J. Peach galaxy listing in *Monthly Notices of the R.A.S.*, 181:323 (1977). **Column 4:** catalogue number in the *Revised New General Catalogue* by J. Sulentic and W. Tifft, Univ. of Arizona Press, 1973, the *Index Catalogues*, and additional listings. **Column 5:** coordinates for epoch 2000.0. **Colunn 6:** V magnitudes unless superscript B, meaning blue magnitude. **Column 7:** mean surface brightness in magnitudes per square arcminute. **Column 8:** angular size in arcminutes.

Abell 1656

Photo by Thomas L. Dessert (10-inch f/6 reflector, 103a-F Film, 25-minute exposure).

75°, is fairly faint and moderately concentrated, but only 10″ diameter.

NGC 4869 is a brighter galaxy located 2′ southwest of the magnitude 12.5 star a similar distance southwest of NGC 4874. In the 12½-inch it is larger and brighter than NGC 4872, and has a magnitude 13.5 star 17″ northwest of its center. The halo is 30″ across, just reaching this star, and is not strongly concentrated.

The supergiant elliptical NGC 4889 is distinctly brighter than NGC 4874 in the 5-inch. The light is well concentrated to a faint stellar nucleus. The 1.5′ halo seems elongated east/west, but a magnitude 13.5 star 1′ southeast makes this impression uncertain. The elongation is nevertheless more certain in larger apertures. With the 12½-inch, however, the center is more broadly concentrated. At 225x I discerned a mottled area in the west side: this is the tiny galaxy **GMP 2976**, only 28″ from the center.

Opposite NGC 4889 from the magnitude 13.5 star is **NGC 4886**, typical of the tiny objects scattered throughout the cluster. A little brighter is **NGC 4898**, a binary galaxy 2.5′ southeast of NGC 4889. I noted in the 12½-inch that this object has a fairly high surface brightness. The 25″x15″ patch is elongated in p.a. 70°, and seems to have little surrounding halo. I could not definitely detect the fainter nearby **NGC 4894**, nor separate the two components of NGC 4898 (combined V=13.4).

Spread around the two central ellipticals are many more galaxies. My attention was drawn to several pairs and strings of objects. Well away from the center to the southwest, **NGC 4839** was not difficult to view in the 5-inch, lying between a magnitude 12 star 2.6′ northeast and a magnitude 13.5 star 2.3′ southwest. The small diffuse glow is weakly brighter across the center. It stands out as one of the larger cluster galaxies with the 12½-inch, reaching to 1.1′x0.7′ in p.a. 60°. The halo has a flat brightness profile in the outer regions, then becomes suddenly brighter across the core, which contains a faint but distinct stellar nucleus. Photographs suggest that the elongation observed here may be due to a faint galaxy (**GMP 4943**) superimposed on the southwest side 25″ from center, and a magnitude 15 star in the northeast side. The close pair **NGC 4842AB** lies only 2.6′ of NGC 4839.

Toward the northwest side of the cluster is another close pair of galaxies, **NGC 4841AB**. Only 30″ apart, these two objects are in contact in the 12½-inch. The southwestern component, NGC 4841A, is the larger and brighter of the two. At 255x it is about 40″ in diameter, slightly elongated east-west, with a strong, even concentration to a conspicuous 8″ core. The northeastern component is similarly concentrated, except that the core is a bit more so, but smaller (substellar), and thus appears fainter than that of NGC 4841A.

NGC 4848 is distinctly elongated with the 12½-inch (it is one of the few spirals in the cluster), reaching to 48″x30″ in p.a. 60°. The lenticular halo has a moderate, even concentration toward the center. **NGC 4860** is of similar size and brightness, but nearly circular. Its light is evenly concentrated to an indistinct core and faint nucleus. I could

The Center of Abell 1656

Photo by John B. Priser using the 61-inch astrometric reflector at the U.S. Naval Observatory's Flagstaff station. U.S.N.O. Photo.

Abell 1367

Photo by Brian Skiff using the 13-inch astrometric "Pluto camera" at Lowell Observatory. See table on page 104 for galaxy identifications.

just make out **NGC 4858** at high power using averted vision.

NGC 4865 lies only 2.8′ northwest of the bright 7th magnitude star. The galaxy is still easy for the 12½-inch. The concentrated halo is about 30″ long, extending east-southeast/west-southwest. Its faint companion, **GMP 3656**, 1.4′ west-southwest, is 20″ across, a diffuse unconcentrated glow difficult for the 12½-inch telescope.

In the field immediately west of NGC 4874 are three small objects. **NGC 4864** is the easiest: in the 12½-inch it is about 30″ across and somewhat elongated toward its companion **NGC 4867**, in about p.a. 120°. The halo is well concentrated to a substellar core. Tiny NGC 4867, about 35″ away, is a sharply concentrated substellar spot clearly separable from NGC 4864. Nearby **IC 3955** forms the western corner of a triangle including a magnitude 15 star 45″ east-northeast, and the galaxy pair NGC 4864/NGC 4867. It is very small, only 10″ across, with a very faint stellar nucleus.

Riding just above the two big guys near the center of the cluster is NGC 4883. In the 12½-inch it is a diffuse, moderately-low surface brightness patch 15″ across. Farther north, **NGC 4881** is the first of three brighter galaxies in a line. It appears as a 20″ circular glow, well concentrated to a sharp, conspicuous stellar nucleus. Next is the spindle **NGC 4895**. It is conspicuous in the field by its shape, a 25″ x 10″ oval elongated in p.a. 150°. The halo has a moderately-high surface brightness and is well concentrated toward the center. Last in the string of three is **NGC 4907**, a moderately faint object for the 12½-inch. The 30″ halo is roughly circular, and weakly concentrated to a faint but distinct stellar nucleus. A magnitude 13 star lies 40″ southwest of center.

Southwest of the center of the cluster is a group of bright spirals. **NGC 4911** is visible in the 5-inch refractor at 60x, forming the southeastern member of a string including two magnitude 12.5 stars within 4′ northwest. With the 12½-inch it is circular, 1.3′ diameter, and broadly concentrated. The companion **GMP 2393** appears to be on the edge of the halo (35″ from center). It is very faint but well-defined, yet unconcentrated toward the center. A magnitude 14.5 star is visible off the halo of NGC 4911, 45″ southeast of its center.

NGC 4919 is a fainter object also with a companion. About 15″ across, it is well concentrated to a stellar nucleus. **GMP 2252**, 1.9′ away in p.a. 290°, is a very faint and concentrated substellar spot. The brightest of the group of spirals is **NGC 4921**, visible in the 5-inch as an unconcentrated 2′ patch with a magnitude 13 star about 3′ northeast. In the 12½-inch the unconcentrated low surface brightness halo is elongated in p.a. 120°, reaching 1.2′x0.8′. The core is brighter, however, and a stellar nucleus is clearly visible. **NGC 4923**, 2.6′ southeast, seems only a little fainter than NGC 4921, but it is much smaller. The diffuse 30″ halo is moderately concentrated to a bright substellar center.

Abell 1367

NGC 3842 is the brightest galaxy in this cluster. Unlike the Coma *lucidae*, this object is of normal luminosity, not a dominating supergiant. In the 5-inch

The Center of Abell 1367

Photo by John B. Priser using the 61-inch astrometric reflector at the U.S. Naval Observatory's Flagstaff station. U.S.N.O. Photo.

Galaxies in Abell 1367
11h44m17.5s, +19°50′30″ (2000)

Type	GP	Object	RA (2000) Dec	Mag.	Size
S0 pec	1522	NGC 3816	11h41m49s +20°06′21″	13.6	2.2′ x 1.3′
(R)SAB(s)ab	1451	NGC 3821	11h42m10s +20°19′03″	13.8	1.6′ x 1.3′
S	1388	Zw 97.72	11h42m46s +20°02′03″	15.0	1.3′ x 0.6′
S0	1290	Zw 97.74	11h42m60s +20°05′15″	15.4	0.8′
Sb	1270	UGC 6680	11h43m03s +19°39′03″	15.5	1.3′ x 0.8′
S0	1169	IC 2951	11h43m26s +19°45′03″	15.0	1.5′ x 0.8′
S	1129	Zw 97.83	11h43m32s +19°37′45″	15.2	1.2′ x 1.0′
E	1083	Zw 97.85	11h43m38s +19°36′21″	15.7	0.2′ x 0.2′
S0	1030	Zw 127.40	11h43m45s +20°16′27″	15.3	—
S0	1015	Zw 97.86	11h43m48s +20°02′27″	15.7	0.6′
Sm	993	UGC 6697	11h43m51s +19°57′57″	14.3	1.7′ x 1.4′
E	949	NGC 3837	11h43m58s +19°53′39″	14.2	1.0′ x 0.9′
E	946	Zw 97.90a	11h43m58s +19°57′27″	17.3	0.2′ x 0.2′
QSO	945	—	11h43m58s +19°56′51″	19.1	0.3′
E	941	Zw 97.90b	11h43m59s +19°57′15″	15.8	0.3′ x 0.3′
Sa	926	NGC 3840	11h43m60s +20°04′39″	14.7	1.2′ x 0.7′
S0	924	Zw 97.92	11h44m01s +19°46′51″	15.5	1.3′ x 0.4′
S0/a	915	NGC 3844	11h44m02s +20°01′45″	14.9	1.5′ x 0.4′
E	900	NGC 3841	11h44m03s +19°58′21″	15.0	1.6′
E3	899	NGC 3842	11h44m03s +19°57′03″	13.3	1.2′ x 1.0′
S	874	NGC 3845	11h44m07s +19°59′51″	15.1	0.8′ x 0.4′
QSO	872	—	11h44m07s +19°57′39″	18.9	0.3′
E	857	Zw 97.99	11h44m09s +19°44′20″	15.7	1.2′
E	800	Zw 97.102a	11h44m18s +20°13′02″	15.6	0.8′ x 0.5′
S	792	Zw 97.102b	11h44m19s +20°13′26″	15.2	0.4′ x 0.3′
pec	780	Zw 97.101	11h44m20s +19°50′32″	15.3	0.6′ x 0.5′
E	771	NGC 3851	11h44m22s +19°58′50″	15.2	0.3′ x 0.3′
S pec	768	Zw 97.110	11h44m22s +19°49′32″	15.5	1.0′
S0	692	Zw 97.112	11h44m32s +20°04′38″	14.9	1.3′ x 0.3′
Sc	616	UGC 6719	11h44m47s +20°07′32″	14.6	1.3′ x 0.9′
Sab	605	NGC 3860	11h44m49s +19°47′44″	14.5	1.3′ x 0.7′
S	603	NGC 3857	11h44m50s +19°32′02″	15.1	1.0′ x 0.5′
S pec	591	NGC 3859	11h44m52s +19°27′20″	14.9	1.1′ x 0.3′
S	579	Zw 97.125	11h44m55s +19°46′38″	15.6	0.7′ x 0.3′
S0	573	Zw 97.124	11h44m57s +19°43′56″	15.3	0.8′
(R′)SAB(r)b	532	NGC 3861	11h45m04s +19°58′26″	14.0	2.4′ x 1.5′
E	531	IC 2955	11h45m04s +19°37′14″	15.2	0.4′ x 04′
E0	521	NGC 3862	11h45m05s +19°36′26″	13.7	1.6′ x 1.6′
S	509	UGC 6725	11h45m06s +20°26′20″	14.5	1.8′ x 1.5′
S	502	—	11h45m06s +19°58′02″	15.6	0.8′ x 0.3′
E	459	Zw 97.131	11h45m15s +19°50′44″	15.1	0.4′ x 0.4′
S	427	Zw 127.47	11h45m24s +20°19′32″	14.6	0.5′ x 0.4′
S	398	NGC 3867	11h45m30s +19°24′02″	14.6	1.7′ x 0.7′
S0	397	NGC 3868	11h45m30s +19°26′44″	14.8	1.2′ x 0.4′
E	319	NGC 3873	11h45m46s +19°46′26″	14.2	1.1′ x 1.0′
S0	319	NGC 3875	11h45m50s +19°46′02″	14.8	1.1′ x 0.3′
SA(r)0/a	217	NGC 3884	11h46m13s +20°23′26″	14.0	1.9′ x 1.5′

Column 1: revised Hubble galaxy types from the *Second Reference Catalogue of Bright Galaxies*, by G. and A. de Vaucouleurs and H. Corwin, Univ. of Texas Press, 1976. **Column 2**: catalogue numbers from J. Godwin and J. Peach galaxy listing in *Monthly Notices of the R.A.S.*, 181:323 (1977). **Column 3**: catalogue number in the *Revised New General Catalogue* by J. Sulentic and W. Tifft, Univ. of Arizona Press, 1973; the *Index Catalogues*, Nilson's *Uppsala General Catalogue*, or Zwicky's *Catalogue of Galaxies and Clusters of Galaxies*. **Column 4**: coordinates for epoch 2000.0. **Column 5**: photographic B (blue) magnitudes, primarily from Zwicky's *CGCG*. **Column 6**: angular size in arcminutes.

refractor it appears as a conspicuous diffuse spot 2.7′ southwest of a magnitude 11 star. At high power the 12½-inch shows it in the midst of a field crowded with fainter galaxies. The 50″ halo contains a small indistinct core and stellar nucleus. **GP 941**, a tiny elliptical companion at 1.1′ in p.a. 280°, is visible in the 12½-inch as a concentrated substellar spot with a magnitude 15 star 35″ northwest of it. I found no trace of an even smaller companion to this galaxy, **GP 946**. **NGC 3841** lies 1.3′ north of NGC 3842. The 12½-inch shows a conspicuous well-concentrated glow here, typical of the fainter E-type galaxies in the cluster.

A few arcminutes to the west is perhaps the most striking object in the cluster, **UGC 6697**. Photographs suggest that it is an edge-on late-type spiral like NGC 55 in Sculptor, but some distortion by NGC 3842 is probably also evident. The 12½-inch shows the 1.2′x0.15′ spindle elongated in p.a. 135°. The brightness along the major axis is approximately uniform except at the very tips.

In the same field to the south, **NGC 3867** seems only a little fainter than NGC 3842. In the 12½-inch the 30″ halo is not quite as well concentrated as the brighter galaxy. To the north, **NGC 3844** and **NGC 3840** appear similar in the 12½-inch. The first is slightly elongated northeast/southwest, while NGC 3840 is a bit larger and better concentrated than NGC 3844. A very faint "anonymous" galaxy, **GP 1015**, could be seen only with difficulty.

Off at the northwest edge of the cluster is **NGC 3821**, located about 3′ east of a magnitude 11 star, and visible in the 5-inch as a hazy spot. At first sight in the 12½-inch it looked like a magnitude 13 double star: the galaxy is the northeastern component, 16″ away from a star of similar magnitude. A very faint halo of the galaxy just envelopes the star, making it about 30″ across. The light is somewhat less concentrated than the star, and a small granular core is visible in the center.

GP 792 and **GP 800** make a difficult pair of galaxies for the 12½-inch, lying only 27″ apart. The two patches are in contact at 225x, the southern being substellar, while the companion is visible only intermittently with averted vision.

One of the larger objects in Abell 1367 is **NGC 3860**, easily visible in the 5-inch refractor at 60x. The 12½-inch shows that the broadly concentrated halo is elongated slightly northeast/southwest. **UGC 6719**, about 20′ north, is practically a clone of NGC 3860.

Another relatively large object is **NGC 3861**, which has a fainter galaxy apparently projected on it — **GP 502**. They are visible as a single diffuse patch in the 5-inch. The glow is nearly 1′ across in the 12½-inch, a plain fuzz with a small sharp center. The companion (50″ from the center in p.a. 115°) is not distinguishable as a separate object, but merely adds to the fuzz on that side of the brighter object.

NGC 3862 is about the second brightest galaxy in the cluster, and it too has a close companion. The brighter object is an easy target for the 5-inch, while the 12½-inch shows both NGC 3862 and **IC 2955**, 55″ away in p.a. 345°. NGC 3862 is 45″ diameter with a moderately strong but azonal brightness profile; a stellar nucleus is occasionally visible. IC 2955 is fainter but not difficult to see, and clearly separated from NGC 3862. It is only 20″ across and rather more diffuse.

A final pair of galaxies is **NGC 3873** and **NGC 3875**, separated by about 50′. Both are visible at low power in a 10-inch telescope. At high power, the 12½-inch showed NGC 3873 as clearly nebulous, and is elongated a bit southeast/northwest. NGC 3875 is similar to a field star nearby to the southeast, but becomes a well-concentrated galaxy with attention.

* * *

I drew the information in the tables from a wide array of sources. For both clusters, identifying numbers are given for every object according to the photometric surveys by Godwin and collaborators. NGC or IC identifiers are given for most galaxies. The GMP numbers in Coma refer to the giant 6700-entry catalogue of Godwin, Metcalfe, and Peach. The GP numbers (from an earlier study by Godwin and Peach) provide a cross-reference to the discussion in the Webb Society Handbook, volume 5. Amazingly, no NGC or IC ambiguities are present in Coma! In Abell 1367, Nilson (in the Uppsala General Catalogue, UGC) calls N3841 what all other catalogues designate NGC 3845, and I've followed the majority opinion here. For Abell 1367, objects without an NGC or IC number are given designations where possible from the UGC (Uxxxx) or from Zwicky's Catalogue of Galaxies and Clusters of Galaxies (CGCG) in the form: ZwXX.XXX.

Positions were derived from the X,Y coordinates given in the GMP or GP catalogues, and are nominally accurate to two arcseconds. The additional precision is quite justified here, particularly for the Coma cluster, where many objects have the same right ascension to one second of arc.

In the Coma cluster, most of the galaxy types are from the Second Reference Catalogue of Bright Galaxies (RC2) by Gerard de Vaucouleurs and collaborators. In Abell 1367 they are a hodgepodge from the RC2, UGC, plus a few of my own from inspection of the POSS and large-scale plates. Dimensions are from the RC2 or from the UGC and MCG (Morphological Catalogue of Galaxies, by Boris Vorontsov-Velyaminov and collaborators) converted to the system of the RC2. Single figures are diameters from the GMP/GP catalogues, not on the standard system, and are "ballpark" figures listed only as a general guide.

Total V magnitudes are given for almost all Coma galaxies; about half are from the RC2. Godwin and Peach give V_{25} (brightness within the magnitude 25 per square arcsecond isophote) for many additional galaxies. These can be approximately corrected to V_t (total magnitude) using information in the RC2. After doing so, I compared the magnitudes for 28 objects having magnitudes listed in the RC2. The mean difference (a few hundredths of a magnitude) is negligibly small, though the scatter is rather larger than that given for the same galaxies in the RC2. Thus I have used the RC2 values when available, and the converted GP magnitudes otherwise (with no special mark). When dimensions are also available, the mean surface brightness magnitude can be calculated, and this is listed in the last column.

The GP magnitudes in Abell 1367 cannot be transformed yet to the standard system because there is little published UBV photometry for these galaxies. Until more definitive results are available, I'll stick with Zwicky's CGCG photographic magnitudes except for objects not listed by him, which come directly from GP. It's important to remember that the Abell 1367 magnitudes are photographic "blue" values, whereas the Coma data are "visual."

Brian Skiff is an astronomer at Lowell Observatory in Flagstaff, Arizona. His first book on observing will be published by Cambridge University Press in 1985. His last article in DEEP SKY was "All About M31," in DSM #8.

Selected Bibliography

Listed below are a few basic articles relevant to the Coma and Abell 1367 clusters, and to galaxy clusters generally. The review articles in the book-series *Annual Reviews of Astronomy and Astrophysics* are emphasized: these are written by leaders in the field, are thorough, and give exhaustive references to more specialized literature. They are also more likely to be found in city and college libraries than even the most common professional journals.

George O. Abell. 1975. "Clusters of Galaxies," in *Galaxies and the Universe*, volume 9 of "Stars and Stellar Systems," Gerard P. Kuiper, ed. University of Chicago Press, Chicago. (The best introductory review for amateurs.)

Neta A. Bahcall. 1977. "Clusters of Galaxies," in *Annual Reviews of Astronomy and Astrophysics*, **15**:505. (Thorough, up-to-date review; but not light reading.)

Alan Dressler. 1984. "The Evolution of Galaxies in Clusters," in *Annual Reviews of Astronomy and Astrophysics*, **22**:185. (Tour-de-force article on the next step: are galaxies in rich clusters different? And what happens to them?)

J.G. Godwin, N. Metcalfe, and J.V. Peach. 1983. "The Coma Cluster. I. A Catalogue of Magnitudes, Colours, Ellipticities, and Position Angles for 6724 Galaxies in the Field of the Coma Cluster." *Monthly Notices of the R.A.S.*, **202**:113 and microfiches.

J.G. Godwin, and J.V. Peach. 1977. "Studies of Rich Clusters of Galaxies. IV. Photometry of the Coma Cluster." *Monthly Notices of the R.A.S.*, **181**:323. (V_{25} magnitudes for 923 galaxies.)

J.G. Godwin, and J.V. Peach. 1982. "Photometry of the Cluster of Galaxies A1367." *Monthly Notices of the R.A.S.*, **200**:733 + microfiche. (Blue and red magnitudes for 1561 galaxies.)

Stephen A. Gregory, and Laird A. Thompson. 1978. "The Coma/A1367 Supercluster and Its Environs." *Astrophys. Journ.*, **222**:784. (Redshift survey showing the existence of this supercluster.)

Jan H. Oort. 1983. "Superclusters." *Annual Reviews of Astronomy and Astrophysics*, **21**:373. (Observationally-oriented review.)

C.D. Shane. 1975. "Distribution of Galaxies," in *Galaxies and the Universe*, volume 9 of "Stars and Stellar Systems," Gerard P. Kuiper, ed. University of Chicago Press, Chicago. (Summary and maps of the Lick survey and what it could mean.)

A Night of Galaxies Near M13

by Robert Bunge

My smarts tell me that because you're reading *Deep Sky* magazine, you've probably stared at M13, the Hercules Cluster, a few times in your life. If you own a large telescope or have access to one, chances are you've seen the little galaxy just northwest of M13, designated NGC 6207. This distant spiral makes a fascinating aside to watching the big ball of stars because it is many times more distant.

However, chances are you haven't seen the really tough objects lying near M13. A close-up examination of the Palomar Observatory *Sky Survey* (*POSS*) plate containing M13 reveals a wealth of faint galaxies scattered within a 4° radius of the great cluster. While some of these galaxies are NGC and IC objects, most of the galaxies are "anonymous," which means they are not listed in the familiar, basic sources of deep-sky data. However, each of these remote galaxies is listed in one or more of three monster galaxy catalogs.

Invaluable sourcebooks for deep-sky researchers, the *Uppsala General Catalogue of Galaxies* (*UGC*), the *Morphological Catalogue of Galaxies* (*MCG*) and the *Catalogue of Galaxies and Clusters of Galaxies* (*CGCG*) together provide information on about 72,000 galaxies. The objects range from the brightest galaxies in the sky all the way down to objects with photographic magnitudes of 15.7.

One day I decided to investigate these faint galaxies near M13. Using the *POSS* plate as a guide, I plotted the galaxies on chart 114 of *Uranometria 2000.0*, with the thought of pushing my favorite deep-sky telescope to its limit. On the next favorable night, I took the chart out to the 31-inch f/7 reflector at Warren Rupp Observatory near Mansfield, Ohio. Each summer I spend great amounts of time showing M13 to visitors — nearly 4,000 people peeked at it in 1988 — and now I wanted to find something a little less mundane that I could show them along with the great cluster.

In the 31-inch scope at 170x, the M13 field appears covered with stars that blanket the view edge-to-edge. The cluster is resolved to the very center of its core. It's easy to follow lines of stars and see dark lanes and voids that seem to hover in front of the cluster.

After enjoying the view of M13, I slowly turned the telescope to **NGC 6207**, located two-and-a-half fields of view to

The Hercules Cluster, M13, is one of the most watched deep-sky objects in the summer sky. Yet several galaxies unknown to most observers lie nearby the great cluster. Photo by Martin C. Germano (6-inch f/8 refractor, hypered Tech Pan film, 30-minute exposure).

the northwest. Because this galaxy has a high surface brightness, it is easily visible with direct vision and shows mottling across its cigar-shaped surface. Because it is so easy to see, I have often shown NGC 6207 on public nights, if visitors seem to be interested in seeing a galaxy.

But a surprise lies halfway between the great cluster and its little companion. Although it is catalogued as an Sb-type galaxy with a magnitude of 12.9, **IC 4617** is not plotted on *Uranometria*. Although this tiny galaxy is relatively bright, it is an unusual challenge, even with a 31-inch telescope. IC 4617 appears to lie sandwiched between two stars that belong to a small parallelogram of stars. In the 31-inch scope I nearly always need averted vision to see IC 4617's tiny, fuzzy, featureless oval. If IC 4617 eludes you, not to worry. Other galaxies near M13 are far easier targets for small backyard telescopes. This is evidenced, in fact, by glancing at the *Uranometria* chart.

About 50' southeast of M13, *Uranometria* shows a small group of galaxies including **NGC 6194**, **NGC 6196**, **NGC 6197**, **NGC 6199**, and **IC 4614**. Unfortunately, a positional error in the *Revised New General Catalogue* caused these galaxies to be mislabeled on *Uranometria*. When I aimed the 31-inch scope toward the area shown on *Uranometria*, I observed three galaxies aligned north-south in a neat line. After making a drawing, I wandered off to other nearby objects.

Later I discovered Steve Gottlieb's letter in *Deep Sky* #5 (Winter 1983), and I read that

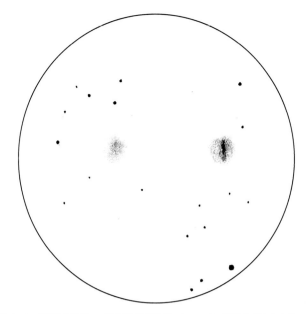

Galaxies UGC 10567 and UGC 10568 lie due east of M13. All sketches in this article are by Robert Bunge using the 31-inch telescope (at 200x) at Warren Rupp Observatory.

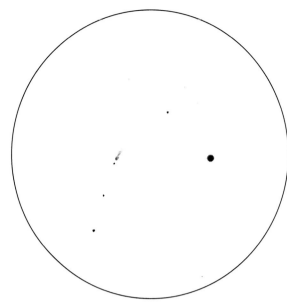

Sixteenth-magnitude galaxy Zwicky 197.016 lies in the same field as the bright star SAO 65578.

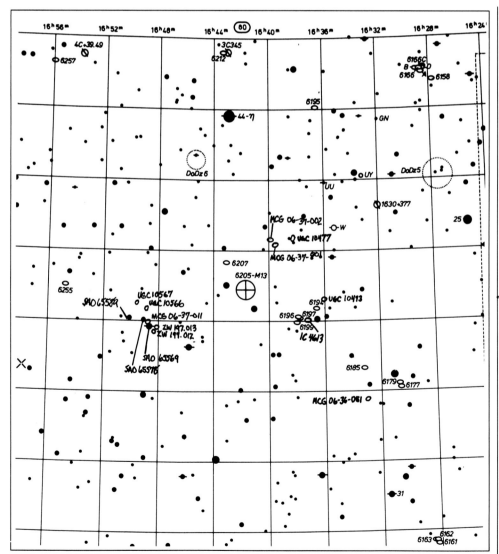

Chart 114 from Uranometria 2000.0, volume 1, shows the area of the Hercules Cluster. In addition to the bright galaxy NGC 6207, thirteen other galaxy fields lie nearby. Copyright © 1987 Willmann-Bell, Inc.

the line of three galaxies is formed by NGC 6197, NGC 6196, and IC 4614. The center galaxy, NGC 6197, is the brightest. In the 31-inch scope it was easy to see NGC 6197 as a tiny, 13th-magnitude glow. NGC 6196, a smaller, fainter blob, lies 3' south of NGC 6197. NGC 6196 showed a bright nucleus with an elongated disk around it, appearing about 14th magnitude overall.

Three arcminutes north of NGC 6197 is IC 4614. At 200x, this galaxy appeared to have a bright center core with a 2" halo surrounding it. It was noticeably fainter than NGC 6197. In his letter, Gottlieb described how Harold Corwin of the University of Texas at Austin had discovered that NGC 6199 is not a galaxy at all but a 14th-magnitude star. Sixteen arcminutes westnorthwest of the line of galaxies is NGC 6194. I logged this object as an easy 13th-magnitude patch of light showing some mottling at 200x.

On the northwest side of M13 lies a galaxy that is plotted but unidentified on *Uranometria*. The first time I centered the big reflector on this area, I couldn't help noticing two galaxies in the field. On close inspection, it dawned on me that a fair amount of homework would be required to identify them. So I resorted to the monster catalogs and the

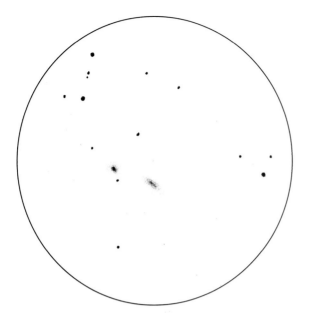 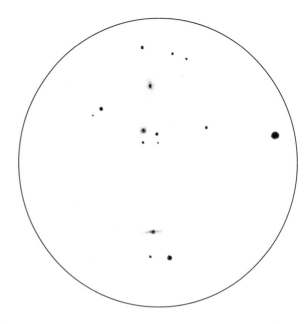

Dimly glowing at magnitudes 14.5 and 15.0, galaxies MCG+06-37-001 and MCG+06-37-002 form a pair even in a high-power eyepiece.

The NGC 6196 galaxy group lies 1° southwest of M13, and contains three galaxies brighter than 15th magnitude.

POSS. The plotted galaxy appears to be **MCG +06-37-002**. Listed at magnitude 14.5, this object appeared round with a stellar core. A 15th-magnitude star lay a few arcseconds south of the core, just outside of the edge of MCG +06-37-002's 3"-diameter glow. Twelve arcseconds east and a few arcseconds south of this galaxy I found the magnitude 15.2 galaxy **MCG +06-37-001**. This galaxy was substantially more difficult to see than MCG +06-37-002, but did offer a 5" elongation that I clearly observed at 200x. Under superb seeing, I would guess that telescopes much smaller than the 31-inch should show this feature.

Up to this point, I had observed the galaxies using only *Uranometria*. During a check for really faint galaxies, I noticed the entire area surrounding M13 was studded with anonymous galaxies. In fact I checked the *POSS* plate with a magnifying glass and noticed many unplotted galaxies just within reach of the plates. Some of these tiny galaxies — far beyond visual reach — appeared to be lodged within the boundaries of M13 itself.

A few nights after viewing the *POSS* plate I arrived at the observatory with Columbus telescope maker Bill Burton. We were both keen on making a systematic search for anony-

The most familiar galaxy near M13 is the bright Sb-type spiral NGC 6207, visible in the same low-power field as the great cluster. Photo by Al Lilge (12.5-inch f/8 Pro-Scope, 098 film, 120-minute exposure).

Faint Galaxies near M13

Galaxy	R.A.(2000.0)	Dec.	Mag.	Type	Notes
UGC 10477	16h37.4m	+37°16'	15.4	Sb	Not found
MCG+06-37-001	16h39.8m	+37°10'	15.0		
MCG+06-37-002	16h40.1m	+37°11"	14.5		
IC 4617	16h42.1m	+36°41'		Sb	
UGC 10544	16h46.7m	+36°05'	15.4	Sa	=Zwicky 197.009
Zwicky 197.012	16h48.3m	+35°50'	15.3		
Zwicky 197.013	16h48.3m	+35°54'	15.3		
MCG+06-37-011	16h48.7m	+35°55'	15.5		=Zwicky 197.014
UGC 10566	16h48.9m	+36°11'	15.3	S	=MCG+06-37-012
UGC 10567	16h49.1m	+36°13'	15.2		=MCG+06-37-013
Zwicky 197.016	16h49.4m	+36°03'	15.7		

Notes: The *UGC* gives sizes for UGC 10544 (1.0" by 0.5"), UGC 10566 (2.0" by 0.4"), and UGC 10567 (1.1" by 0.6"). The double star near MCG+06-37-011 is SAO 65569 (V=7.3). The next bright star east (near Zwicky 197.016) is SAO 65578 (V=8.4). The next star to the east is SAO 65589 (V=8.0).

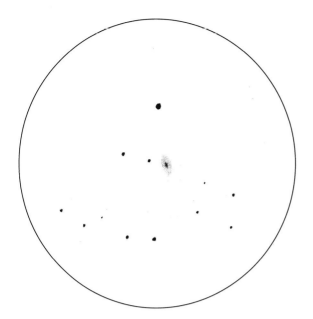

UGC 10544 is a 15th-magnitude Sa-type spiral with a bright, starlike nucleus.

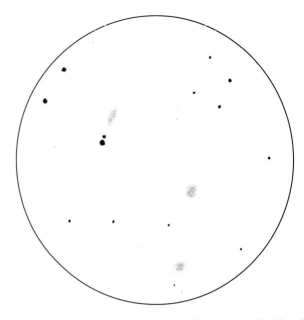

A complex and interesting field southeast of M13 contains two Zwicky galaxies and MCG+06-37-011, which lies adjacent to the double star SAO 65569.

mous galaxies in the region. No other use was scheduled for the 31-inch scope that night. M13 was placed high in the western sky. After looking at M13, I hopped over to the two MCG galaxies. Then, using the scope's 5-inch finder, I moved over to the position of **UGC 10447**, a faint, edge-on galaxy located between two bright stars.

I knew that I had the right field because the two stars were on opposite ends of the field of view at 200x. However, neither Bill nor I could see the galaxy, even at extremely high magnifications. The *POSS* shows UGC 10447, and the *UGC* lists its magnitude as 15.4. Our only suspected reason for not seeing the galaxy was overpowering glare from the two bright stars.

With that disappointing start, I moved the telescope back to M13 and continued to the position of **UGC 10544**, a little over 1° to the east-southeast. This galaxy was plainly visible with averted vision. It had a well-defined core surrounded by a milky, oval glow. I estimated its size at about 3" by 2". UGC 10544 must have a fairly high surface brightness, because we saw it rather easily and the *UGC* lists it at magnitude 15.4. Incidentally, many so-called anonymous galaxies are listed in more than one of the monster catalogs. UGC 10544 is also known as MCG +06-37-009.

About 3' east and a few arcminutes south of UGC 10544 lies a triangle consisting of two 8th-magnitude stars and a 7th-magnitude star shown as a double on *Uranometria*. Inspection of the *POSS* plate revealed a wealth of faint galaxies surrounding these stars. So with a fair degree of anticipation, I pointed the scope at the double.

I had a 28mm Pretoria eyepiece attached to the scope, yielding 200x. Peering into the eyepiece, I saw the star was indeed double, with the secondary lying to the northwest. Two galaxies were immediately apparent in the field. After carefully scanning the field, Bill pointed out a third galaxy close to the secondary. This object was significantly dimmer than the other two.

Later we identified this galaxy as **MCG +06-37-011**. This galaxy appeared to be about 16th magnitude — not an easy find even for the 31-inch scope. MCG +06-37-011 appeared spread out with an extension on its northwestern side. The galaxy did not appear to have a central brightening.

Some 24" west and 1" south of MCG +06-37-011 is the more impressive **Zwicky 197.013**, the best of the three galaxies surrounding the double star. Glowing at magnitude 15.3, this object clearly had a higher surface brightness than MCG +06-37-011. Indeed,

Zwicky 197.013 was round and considerably smaller than the MCG galaxy. Located just 4" east of Zwicky 137.013 was **Zwicky 137.012**. This galaxy appeared larger and dimmer than its close neighbor.

Bumping the scope a field diameter to the northeast brought one of the 8th-magnitude stars, SAO 65578, into the field along with **Zwicky 197.016**, a magnitude 15.7 galaxy located northeast of the star. This galaxy at first appeared double because of a faint star of about the same brightness lying adjacent to the galaxy's center.

However, Zwicky 197.016 is an edge-on galaxy with a foreground star. After studying the galaxy carefully with averted vision, I could detect an elongation oriented north-south. Zwicky 197.016 appeared to be slightly larger toward the star. Just to be sure, I checked the *POSS* plate again and found a star in the observed position. I hadn't discovered a distant supernova.

The night had progressed more quickly than we would have liked. The once-high M13 was now sinking into the glow of Mansfield. I moved the scope north-northwest to the position of **UGC 10566**. This magnitude 15.3 galaxy was easily visible in the 31-inch scope. Averted vision showed a distinct brightening along a north-south orientation. The *POSS* shows clearly that this galaxy is a barred spiral.

Just 12" east and 2" north of UGC 10566 is **UGC 10567**. As I peered into the eyepiece and complained that this was the toughest galaxy of the night, Bill asked me to look up. Not only was M13 now down into light pollution, but the telescope was partially blocked by the dome slit. Nonetheless, I saw UGC 10567 as a faint, featureless spot, but I suspect it is much more impressive under better conditions.

After a couple of hours of searching, Bill and I had found nine so-called anonymous galaxies near M13. None of the objects was brighter than magnitude 14.5. Although we were using a large telescope, we are confident that most of these objects are visible in much smaller instruments. This should especially be true for observers at better sites than ours. This summer, as M13 shines against the black velvet of the night sky, spend an hour or two exploring the secret wonders scattered quietly about the great ball of stars.

Robert Bunge is an active deep-sky observer who lives in Columbus, Ohio. His last article was "Tracking Down the NGC 4005 Galaxy Group" in the Spring 1989 issue.

Bibliography

Arp, Halton C. *Quasars, Redshifts, and Controversies.* 198 pp., hardcover. Interstellar Media, Berkeley, California, 1987. Semitechnical account of how a respected astronomer has found certain galaxies with anomalous redshifts.

Arp, Halton C., and Barry F. Madore. *A Catalogue of Southern Peculiar Galaxies and Associations.* Two vols., 208 pp., and 1,017 black-on-white photographs. Cambridge University Press, New York, 1987. A marvelous atlas of peculiar galaxies in the southern sky, giving the reader a great appreciation of the variety of galaxy forms.

ASTRONOMY, Kalmbach Publishing Co., Waukesha, Wisconsin. Founded in 1973, this monthly is the largest English-language astronomy periodical. It regularly contains plentiful information about deep-sky objects.

Burnham, Robert, Jr. *Burnham's Celestial Handbook.* Three vols., 2,138 pp., paper. Dover Publications, New York, 1978. This voluminous compilation of deep-sky objects contains many photographs, charts, and tables.

Corwin, Harold R., Jr., Antoinette de Vaucouleurs, and Gerard de Vaucouleurs. *Southern Galaxy Catalogue.* 308 pp., paper. The University of Texas Press, Austin, 1985. This valuable catalog contains fundamental data for 5,841 galaxies south of declination -17°, assembled from studies of photographs made with the U.K. Schmidt telescope.

de Vaucouleurs, Gerard, Antoinette de Vaucouleurs, Harold R. Corwin, Jr., Ronald J. Buta, Georges Paturel, and Pascal Fouque. *Third Reference Catalogue of Bright Galaxies.* Three vols., 2,069 pp., hardcover. Springer-Verlag, New York, 1991. The most valuable modern compilation of data for galaxies, covering 23,024 objects.

Dixon, Robert S., and George Sonneborn, compilers. *A Master List of Nonstellar Optical Astronomical Objects.* 835 pp., hardcover. Ohio State University Press, Columbus, 1980. An all-in-one compilation of more than 185,000 deep-sky objects drawn from 270 catalogues. This list is invaluable for identifying objects.

Dreyer, John Louis Emil. *New General Catalogue of Nebulae and Clusters of Stars (1888). Index Catalogue (1895). Second Index Catalogue (1908).* 378 pp., paper. Royal Astronomical Society, London, 1962. This single volume presents a facsimile reprint of the original basic listings of deep-sky objects compiled by Dreyer.

Eicher, David J. *Beyond the Solar System; 100 best deep-sky objects for amateur astronomers.* 80 pp., paper. AstroMedia, a division of Kalmbach Publishing Co., Waukesha, Wisconsin, 1992. An introduction to the brightest and most spectacular of the sky's nebulae, clusters, and galaxies.

Eicher, David J., and the editors of *Deep Sky* magazine. *Deep Sky Observing with Small Telescopes.* 331 pp., paper. Enslow Publishers, Hillside, New Jersey, 1989. A beginner's manual for observing deep-sky objects with 2-inch to 6-inch telescopes. Contains extensive listings of objects and many photographs and eyepiece sketches made by backyard observers.

Eicher, David J., ed. *Stars and Galaxies; ASTRONOMY's guide to exploring the cosmos.* 200 pp., paper. AstroMedia, a division of Kalmbach Publishing Co., Waukesha, Wisconsin, 1992. Hundreds of photos, sketches, and diagrams supplement forty articles centered on unusually rich regions of sky.

Eicher, David J. *The Universe from Your Backyard.* 188 pp., hardcover. Cambridge University Press and AstroMedia, a division of Kalmbach Publishing Co., New York, 1988. This book is a series of republished "Backyard Astronomer" articles from ASTRONOMY magazine. Included in its coverage are forty-six constellations or groups of constellations and 690 deep-sky objects. A three-color map, eyepiece sketches, and color photographs appear for each constellation.

Ferris, Timothy. *Galaxies.* 191 pp., hardcover. Stewart, Tabori & Chang, New York, 1980. A folio-sized photo essay on galaxies with an engagingly written narrative.

Hartung, E.J. *Astronomical Objects for Southern Telescopes.* 238 pp., hardcover. Cambridge University Press, New York, 1968. Meticulous observing notes by an Australian observer for many deep-sky objects in the Southern Hemisphere.

Hirshfeld, Alan, and Roger W. Sinnott, eds. *Sky Catalogue 2000.0.* Cambridge University Press and Sky Publishing Corp., New York, 1982-1985. Two vols. Vol. 2 (356 pp., hardcover) lists fundamental data for thousands of double and variable stars, 750 open clusters, 150 globular clusters, 238 bright nebulae, 150 dark nebulae, 564 planetary nebulae, 3,116 galaxies, and 297 quasars.

Hodge, Paul. *Atlas of the Andromeda Galaxy.* 79 pp., hardcover. The University of Washington Press, Seattle, 1982. A single-volume set of photographic maps that display all of the individual features in M31.

Hodge, Paul. *Galaxies.* 174 pp., hardcover. Harvard University Press, Cambridge, Massachusetts, 1986. A revision of the classic introduction to galaxy research written by Harlow Shapley.

Jones, Kenneth Glyn. *Messier's Nebulae and Star Clusters.* Second ed., 427 pp., hardcover. Cambridge University Press, New York, 1991. One of England's foremost amateur astronomers presents descriptions and eyepiece drawings for each of the Messier objects.

Jones, Kenneth Glyn. *The Search for the Nebulae.* 84 pp., hardcover. Alpha Academic, Giles, England, 1975. A compendium of the recorded notes about deep-sky objects by several dozen early observers.

Jones, Kenneth Glyn, ed. *The Webb Society Deep Sky Observer's Handbook.* Eight vols., 1,544 pp., paper. Enslow Publishers, Hillside, New Jersey, 1979-1990. Collection of observations of deep-sky objects by amateur astronomers, made with everything from binoculars to large, professional telescopes. Covers double stars (vol. 1), planetary and gaseous nebulae (vol. 2), open and globular clusters (vol. 3), galaxies (vol. 4), clusters of galaxies (vol. 5), anonymous galaxies (vol. 6), Southern Hemisphere objects (vol. 7), and variable stars (vol. 8).

Luginbuhl, Christian B., and Brian A. Skiff. *Observing Handbook and Catalogue of Deep-Sky Objects.* 352 pp., hardcover. Cambridge University Press, New York, 1990. The authoritative single volume for data on deep-sky objects, this book contains information on nearly 2,050 galaxies, nebulae, and clusters.

Mallas, John H., and Evered Kreimer. *The Messier Album.* 216 pp., hardcover. Sky Publishing Corp., Cambridge, Massachusetts, 1978. A photograph, sketch, and brief description for each Messier object.

Mitton, Simon. *Exploring the Galaxies.* 206 pp., paper. Charles Scribner's Sons, New York, 1976. A semitechnical discussion of galaxies and galaxy research.

The **New Cosmos**. 160 pp., paper. Kalmbach Books, Waukesha, Wisconsin, 1992. The best material from ASTRONOMY magazine provides readers with a hard look at the cutting edge of astronomical research, including much on galaxies.

Newton, Jack, and Philip Teece. *The Cambridge Deep-Sky Album.* 126 pp., hardcover. Cambridge University Press and AstroMedia Corp., New York, 1984. This work contains color photographs of all the Messier objects and many NGC objects.

Sandage, Allan. *The Hubble Atlas of Galaxies.* Fifty plates and accompanying text, hardcover. Carnegie Institution of Washington, Washington, D.C., 1961. Magnificent atlas of galaxies containing black-on-white photographs made with the world's finest telescopes.

Sandage, Allan, and John Bedke. *Atlas of Galaxies Useful for Measuring the Cosmological Distance Scale.* 13 pp. + 95 photographic panels, hardcover. National Aeronautics and Space Administration, Washington, 1988. This magnificent volume contains poster-sized black-on-white prints of nearby galaxies.

Sandage, Allan, and Gustav A. Tammann. *A Revised Shapley-Ames Catalog of Bright Galaxies.* 157 pp., hardcover. Carnegie Institution of Washington, Washington, D.C., 1987. Update of a famous study of 1,246 bright galaxies completed by the Harvard astronomers Shapley and Ames in 1932. Includes an atlas of black-on-white photographs of eighty-four galaxies in the survey.

Sinnott, Roger, ed. *NGC 2000.0; the complete New General Catalogue and Index Catalogues of Nebulae and Star Clusters by J.L.E. Dreyer.* 273 pp., paper. Sky Publishing Corp. and Cambridge University Press, Cambridge, Massachusetts, 1988. A computer-generated revision of the *NGC* and *IC* catalogs with revised data for each object listed and coordinates given for equinox 2000.0. A valuable addition to the deep-sky observer's bookshelf.

Sky & Telescope. Sky Publishing Corp., Cambridge, Massachusetts. The oldest astronomy magazine in America, *Sky & Telescope* contains a monthly "Deep Sky Wonders" column written by the experienced observer Walter Scott Houston.

Sulentic, Jack W., and William G. Tifft. *The Revised New General Catalogue of Nonstellar Astronomical Objects.* 383 pp., hardcover. The University of Arizona Press, Tucson, 1973. The *Revised New General Catalogue* is an update of the standard list of bright deep-sky objects compiled by William and John Herschel. It contains positions, magnitude, and brief encoded descriptions for over 7,800 objects.

Thompson, Gregg D., and James T. Bryan, Jr. *The Supernova Search Charts and Handbook.* 134pp., hardcover, + 100 large quarto charts, boxed. Cambridge University Press, New York, 1990. An important set that constitutes the finest easily available resource for supernova hunting.

Tirion, Wil. *Sky Atlas 2000.0.* Twenty-six fold-out folio charts, spiral-bound. Cambridge University Press and Sky Publishing Corp., New York, 1981. A large-scale atlas showing 43,000 stars down to magnitude 8 and 2,500 deep-sky objects in color.

Tirion, Wil, Barry Rappaport, and George Lovi. *Uranometria 2000.0.* Two vols., 473 quarto-sized charts, hardcover. Wilmann-Bell, Inc., Richmond, Virginia, 1987-1988. A minutely detailed, large-scale atlas, *Uranometria 2000.0* shows 332,556 stars down to magnitude 9.5 and many thousands of deep-sky objects.

Vehrenberg, Hans. *Atlas of Deep-Sky Splendors.* Fourth ed., 246 pp., hardcover. Treugesell-Verlag and Sky Publishing Corp., Dusseldorf, 1981. A splendid photographic album containing images of hundreds of deep-sky objects all reproduced at the same scale for easy comparison.

Wray, James D. *The Color Atlas of Galaxies.* 189 pp., hardcover. Cambridge University Press, New York, 1988. A spectacular tour of galactic forms, this McDonald Observatory astronomer's work displays in color over six hundred galaxies.

Index

0818+71, 36, 37
Abell 426, 98
Abell 1367, 84, 89, 99, 102, 103, 104
Abell 1656, 100, 101
Abell 2151, 84, 98
Abell catalogue, 98
Abell, George, 97, 98,105
ADS 3623, 96
Al-Sufi, 90
Andromeda, 9, 15,54
Andromeda Galaxy, 6,22, 30,59,90
Andromeda I, 9, 10, 11
Andromeda I, 57
Andromeda II, 9, 10
Andromeda II, 57
Andromeda III, 9, 10
Andromeda III, 57
Andromeda IV, 9
Andromeda IV, 57
Anemic spiral galaxy, 19
Aquarius Dwarf, 10, 11
Aquarius, 10
Aries, 15, 21
Arizona, 33
Arp, Halton C., 17, 19, 54, 61
Arp 84, 29
Arp 94, 25
Arp 116, 29
Arp 120, 26
Arp 166, 25
Arp 212, 19
Arp 269, 27
Arp 270, 26
Arp 271, 29
Arp 281, 29
Astier, N., 61
Astronomical Journal, 12, 30, 87, 93
Astrophysical Journal Supplement Series, 12
Atlas Coeli, 70
Atlas of Peculiar Galaxies, 19
Atlas of the Andromeda Galaxy, 55, 59, 91, 92, 93
Averted vision, 94
Baade, Walter, 13, 55
Bahcall, Neta A., 105
Barnard, E. E., 9, 82
Barnard's Galaxy, 9, 10
Barred spirals, 15, 17, 25, 89, 96
Bellatrix, 96
Berry, Richard, 58
Beta Andromedae, 17, 20
Beta Canum Venaticorum, 29
Betelgeuse, 96
Big Bang, 92
Big Dipper, 22
Bigourdan, Guillaume, 62, 65
Binoculars, 54
Blackeye Galaxy, 74
Blue supergiants, 54
Book of the Fixed Stars, 90
Boullian, Ishmael, 90
Brownlee, K. Alexander, 21, 23, 69, 74, 76, 78, 80
Bunge, Robert, 62, 67, 106, 107, 109
Burnham's Celestial Handbook, 62
Burton, Bill, 108
Buta, Ronald J., 97
Canada-France-Hawaii telescope, 55
Canes Venatici, 22, 23, 26, 27, 29, 62, 68, 70, 75, 98
Carina, 10
Carina Dwarf, 13
Carlson, Dorothy, 81
Cassiopeia, 9, 13, 15, 21, 58
Catalogue of Galaxies and Clusters of Galaxies, 66, 80, 97, 106
cD galaxies, 84
Cepheid variables, 9, 30, 91
Cepheus, 15
Cetus, 15, 21
Chayer, P., 61
Coe, Steve, 9
Coma Berenices, 15, 22, 25, 26, 70, 75, 84
Coma Cluster of galaxies, 22, 84, 89, 98

Coma-Sculptor Cloud, 30
Coma-Virgo Supercluster, 89
Coombs, Lee C., 56, 58, 73
Corder, Jeff, 77, 81, 84, 87, 88,89
Corona Borealis galaxy cluster, 6, 98
Corwin, Harold G. Jr., 61, 80, 81, 97, 107
Courtes, C., 55, 61
Cowley, A., 61
Crampton, D., 55, 61
CVn I cloud, 68
Da Costa, Gary, 61
Danielson, R. E., 60, 61
Dark adaptation, 6, 37
DDO 8, 10
DDO 42, 36
DDO 53, 36
DDO 155, 10
DDO 165, 36
DDO 199, 10
DDO 208, 10
DDO 210, 10
DDO 216, 10
de Vaucoulers, Antoinette, 61, 97
de Vaucoulers, Gerard, 61, 97, 105
Deep Sky, 13, 17, 21, 29, 61, 81, 89, 93, 99, 105, 106
Deep Sky Monthly, 29
Density waves, 30
Double Cluster, 55
Double stars, 76, 109
Draco Dwarf galaxy, 10, 13, 82
Dressler, Alan, 105
Dreyer, J.L., 62, 64, 65, 66, 70, 79, 80, 81
Dust clouds, 92
Dwarf galaxies, 82
E206-G220, 10
E351-G30, 10
E356-G04, 10
E594-G04, 10
Eicher, Dave, 60
8 Orionis, 96
Elliptical galaxies, 15, 25, 29, 33, 70, 89, 99
Eridanus, 96
Eta Carinae, 37
Fletcher, Bill, 30
Fletcher, Sally, 30
Ford, W. Kent, Jr., 54, 61
Fornax galaxy cluster, 13
Fornax System, 10, 17, 19, 12
Fouqué, Pascal, 97
Freed, Harvey, 9, 33
G35, 94
G70, 94, 95
G73, 95
G78, 94, 95
G87, 94, 95
G96, 94, 95
G119, 94
G124, 94, 95
G134, 94, 95
G150, 94, 95
G156, 94
G172, 94, 95
G176, 94
G213, 94, 95
G226, 95
G233, 94, 95
G234, 95
G244, 95
G250, 94, 95
G263, 95
G279, 95
Galaxies, 6, 9, 14, 22, 91, 96, 98
Gamma Comae Berenices, 84
Gamma Orionis, 96
Germano, Martin C., 13, 16, 17, 19, 20, 33, 36, 55, 58, 73, 74, 75, 82, 83
Globular clusters, 92, 93
GMP 2252, 103
GMP 2393, 103
GMP 2976, 101
GMP 3656, 103
GMP 4943, 101
Godwin, J. G., 105
Goldstein, Alan, 14, 17, 21, 22
Gottlieb, Steven, 77, 81, 96, 97, 106, 107
GP 502, 104

GP 792, 104
GP 800, 104
GP 941, 104
GP 946, 104
GP 1015, 104
GR 8, 10
Gregory, Stephen A., 98, 105
Grus, 2, 15
HII regions, 9, 36, 37, 54, 55, 61, 70, 75, 78, 79
H-alpha filters, 9
Hartwing, E., 90
Harvard University, 98
Helical Nebula, 83
Hercules galaxy cluster, 22, 84, 98, 107
Herschel, John, 15, 62
Herschel, William, 15, 78, 80, 96
Higgins, David, 6, 90, 94, 95
Hodge, Paul W., 6, 13, 37, 61, 91, 92
Holmberg I, 34, 36, 37
Holmberg II, 35, 36, 37
Holmberg, Erik, 30
Horsehead Nebula, 96
Hubble Atlas of Galaxies, The, 36
Hubble flow, 6
Hubble, Edwin, 6, 9, 15, 92, 98
Hydrogen, 78, 93
IC 10, 9, 10, 11, 13, 15, 19
IC 342, 82, 83
IC 742, 88, 89
IC 839, 99
IC 1450, 78
IC 1459, 19, 21
IC 1613, 9, 10, 11, 19, 21,
IC 1727, 17, 21
IC 2574, 36, 37
IC 2951, 104
IC 2955, 104, 105
IC 2968, 88, 89
IC 3356, 19
IC 3943, 99
IC 3946, 99
IC 3947, 99
IC 3949, 99
IC 3955, 99, 103
IC 3957, 99
IC 3959, 99
IC 3960, 99
IC 3961, 75
IC 3963, 99
IC 3973, 99
IC 3976, 99
IC 3998, 99
IC 4011, 99
IC 4012, 99
IC 4021, 99
IC 4026, 99
IC 4040, 99
IC 4041, 99
IC 4042, 99
IC 4045, 99
IC 4051, 99
IC 4182, 72
IC 4257, 66, 67
IC 4263, 66
IC 4277, 62, 66
IC 4278, 66
IC 4282, 66
IC 4614, 106, 107
IC 4617, 106, 108
IC 5152, 10, 11, 19
Index Catalogue, 75
Interacting galaxies, 22, 26
Iota Draconis, 82
Irregular galaxy, 21
Jenkins, Jamey, 58
Kappa Draconis, 82
Keeler, James Edward, 62, 66
Kennicutt, Robert, 61
Kitt Peak National Observatory, 92
Kristian, Jerome, 61
Kuiper, Gerard P., 105
Large Magellanic Cloud, 15, 22, 70, 75
Leach, R., 61
Lenticular galaxies, 19, 86
Leo A, 10
Leo B, 10
Leo I, 10, 11, 13, 82
Leo II, 10, 13
Leo III, 10

Leo Minor, 26, 28,
Leo, 25, 26, 98
Lick Observatory, 62, 98
Light, E. S., 60, 61
Lilge, Al, 108
Ling, Alister, 82, 83
LMC, 10
Local Group System 3, 9, 10, 21, 30, 68
Low surface brightness galaxies, 6, 9
Lowell Observatory, 33, 56, 100, 102
Luginbuhl, Chris, 61
Lundmark, Knut, 9
Lyuty, V. M., 61
M3, 75
M5, 75
M13, 75, 106, 107, 108
M17, 82
M31, 10, 15, 21, 30, 54, 60, 61, 78, 90, 91, 92
M32, 6, 10, 11, 22, 55, 58, 59, 92, 94,
M33, 10, 15, 19, 25, 54, 61, 82, 91
M42, 82, 96
M49, 58
M51, 22, 23, 29, 62, 68, 75, 90
M60, 29
M63, 75
M64, 74
M81 DW A, 36, 37
M81 Group, 19, 30, 34, 36
M82, 33, 34, 36, 37, 70
M87, 58
M93, 68
M94, 68, 70, 75
M101 group, 19, 54, 61, 68, 91
M104, 15
M105, 26
M106, 68, 70, 71
M110, 10, 55
Maffei 1, 6
Magellanic Clouds, 6, 91
Marling, Jack, 97
Maucherat, A. J., 61
Mayer, Simon, 90
MCG +06-37-001, 108
MCG +06-37-002, 108
MCG +06-37-011, 108, 109
MCG +08-24-093, 66
MCG +08-24-097, 66
MCG +08-25-004, 66
MCG +08-25-005, 66
MCG +08-25-006, 62, 64, 66
MCG +08-25-007, 66
MCG +08-25-009, 66
MCG +08-25-010, 65, 66
MCG +08-25-011, 66
MCG +08-25-015, 66
MCG +08-25-017, 65, 66, 67
MCG +08-25-023, 66
MCG +08-25-025, 66
MCG +08-25-029, 66
MCG +08-25-030, 66
Melotte, P.J., 9
Messier, Charles, 17, 29, 70, 78, 98
Metcalfe, N., 105
Milky Way, 6, 13, 15, 22, 30, 54, 70, 78, 91, 92, 95, 96
Mitchell, Larry, 10
Monnet, G., 61
Monthly Notices of the Royal Astronomical Society, 12
Morphological Catalogue of Galaxies, 65, 106
Mould, Jeremy, 61
Mount Wilson Observatory, 91, 98
Mu Cephei, 54
Mundus Jovialis, 90
National Deep Sky Observers Society, 21, 29, 95
Nearby Galaxies Atlas, 30
Nearby Galaxies Catalog, 30
Nebulae, 6, 90
Nebuleuses et D'Amas Stellaires, 62
Neutral hydrogen, 37
New General Catalogue, 12, 62
NGC 16, 19
NGC 55, 17, 19

NGC 127, 19
NGC 128, 19
NGC 130, 19
NGC 147, 6, 10, 11, 19, 21, 57, 58
NGC 151, 19
NGC 185, 57, 58
NGC 205, 6, 10, 11, 21, 54, 55, 58, 59, 95, 95
NGC 206, 55, 59, 92
NGC 221, 10, 54
NGC 224, 10, 54
NGC 247, 6, 15, 16
NGC 253, 15, 19
NGC 300, 19
NGC 404, 17, 19, 20
NGC 470, 19
NGC 474, 19
NGC 488, 19
NGC 598, 10
NGC 672, 17, 19, 21
NGC 681, 15, 19
NGC 718, 19
NGC 750, 25, 1, 19, 21
NGC 751, 25
NGC 770, 15, 19, 21
NGC 772, 15, 19, 21
NGC 891, 9, 15, 19
NGC 925, 17, 19
NGC 985, 19, 21
NGC 1023, 19
NGC 1097, 17, 19
NGC 1156, 19, 21
NGC 1201, 19
NGC 1232, 19
NGC 1275, 19, 21
NGC 1654, 96
NGC 1657, 96
NGC 1661, 96, 97
NGC 1670, 96, 97
NGC 1678, 96, 97
NGC 1682, 96, 97
NGC 1683, 96, 97
NGC 1684, 96, 97
NGC 1685, 96, 97
NGC 1690, 96, 97
NGC 1691, 96, 97
NGC 1709, 96, 97
NGC 1713, 96, 97
NGC 1719, 96, 97
NGC 1729, 96, 97
NGC 1740, 97
NGC 1753, 96, 97
NGC 1762, 96, 97
NGC 1819, 96, 97
NGC 1843, 97
NGC 1875, 97
NGC 1924, 96, 97
NGC 2000.0, 62, 64
NGC 2110, 97
NGC 2119, 96, 97
NGC 2336, 33
NGC 2363, 97
NGC 2366, 19, 35, 36, 37
NGC 2403, 19, 33, 35, 36, 82
NGC 2976, 36, 37
NGC 3031, 36
NGC 3034, 36
NGC 3077, 36, 37
NGC 3166, 25
NGC 3169, 25
NGC 3226, 25
NGC 3227, 25, 26
NGC 3344, 19
NGC 3384, 26
NGC 3389, 26
NGC 3395, 26, 28
NGC 3396, 26, 28
NGC 3816, 104
NGC 3821, 104
NGC 3837, 104
NGC 3840, 104
NGC 3841, 104
NGC 3842, 103, 104
NGC 3844, 104
NGC 3845, 104
NGC 3851, 104
NGC 3857, 104
NGC 3859, 104
NGC 3860, 104
NGC 3861, 104
NGC 3862, 104, 105
NGC 3867, 104
NGC 3868, 104

111

NGC 3873, 104, 105
NGC 3875, 104, 105
NGC 3884, 104
NGC 3929, 88, 89
NGC 3937 galaxy group member #67, 88, 89
NGC 3937 galaxy group member #68, 88, 89
NGC 3937 galaxy group member #70, 88, 89
NGC 3937 galaxy group member #71, 88, 89
NGC 3937 galaxy group member #72, 88, 89
NGC 3937 galaxy group member #73, 89
NGC 3937 galaxy group member #74, 88, 89
NGC 3937 galaxy group member #82, 88, 89
NGC 3937 galaxy group member #83, 88, 89
NGC 3937 galaxy group member #92, 88, 89
NGC 3937 galaxy group member #100, 88, 89
NGC 3937 galaxy group, 88, 89
NGC 3940, 88, 89
NGC 3943, 88, 89
NGC 3946, 88, 89
NGC 3947, 88, 89
NGC 3954, 88, 89
NGC 3987, 86, 87
NGC 3989, 87
NGC 3993, 86, 87, 89
NGC 3997, 86, 87, 89
NGC 3999, 87, 89
NGC 4000, 87, 89
NGC 4005 galaxy group, 86, 87, 89
NGC 4009, 87, 89
NGC 4011, 87, 89
NGC 4015, 87, 89
NGC 4015A, 87
NGC 4015B, 87
NGC 4018, 87, 89
NGC 4021, 87, 89
NGC 4022, 87, 89
NGC 4023, 87, 89
NGC 4111, 68, 70, 71
NGC 4117, 71
NGC 4138, 70, 71
NGC 4143, 70, 71
NGC 4145, 68, 70, 71
NGC 4145A, 70
NGC 4151, 68, 70, 71
NGC 4156, 70, 71
NGC 4163, 71
NGC 4169 group, 84, 89, 86
NGC 4170, 84, 86
NGC 4174, 84, 86
NGC 4175, 84, 86
NGC 4183, 71
NGC 4185, 84
NGC 4190, 71
NGC 4214, 68, 70, 71
NGC 4217, 68, 70, 71
NGC 4220, 70, 71
NGC 4227, 71
NGC 4231, 70, 71
NGC 4232, 70
NGC 4236, 19, 34, 36, 37, 82, 83
NGC 4242, 68, 70, 71
NGC 4244, 68, 70, 71, 73
NGC 4248, 70, 71, 75
NGC 4258, 71
NGC 4281, 10, 87
NGC 4288, 71
NGC 4290, 68
NGC 4298, 25, 26
NGC 4302, 25, 26
NGC 4346, 71
NGC 4357, 71
NGC 4369, 68, 71
NGC 4389, 71
NGC 4395, 19, 68, 70, 71
NGC 4435, 26
NGC 4438, 26
NGC 4449, 68, 70, 71, 73
NGC 4460, 71
NGC 4485, 26, 27, 29, 70, 71, 74
NGC 4490, 26, 27, 29, 70, 71, 74
NGC 4534, 71
NGC 4565, 15
NGC 4618, 68, 70, 71
NGC 4619, 71
NGC 4625, 70, 71
NGC 4627, 26, 29, 70, 71, 73

NGC 4631, 26, 29, 68, 70, 71, 73
NGC 4647, 29
NGC 4656, 70, 71, 73
NGC 4657, 70, 73
NGC 4736, 71
NGC 4752, 29
NGC 4762, 29
NGC 4800, 68, 71
NGC 4839, 99, 101
NGC 4840, 99
NGC 4841A, 99
NGC 4841AB, 101
NGC 4841B, 99
NGC 4842A, 99
NGC 4842B, 99, 101
NGC 4848, 99, 101
NGC 4851, 99
NGC 4853, 99
NGC 4854, 99
NGC 4858, 99, 103
NGC 4859, 99
NGC 4860, 99, 101
NGC 4861, 71, 75
NGC 4864, 99, 103
NGC 4865, 99, 103
NGC 4866, 10
NGC 4867, 99, 103
NGC 4868, 71
NGC 4869, 99, 101
NGC 4871, 99
NGC 4872, 99
NGC 4873, 99
NGC 4874, 99
NGC 4875, 99
NGC 4876, 99
NGC 4881, 99, 103
NGC 4883, 99, 103
NGC 4886, 99, 101
NGC 4889, 99
NGC 4894, 101
NGC 4894A, 99
NGC 4895, 99, 103
NGC 4896, 99
NGC 4898, 101
NGC 4898A, 99
NGC 4898B, 99
NGC 4906, 99
NGC 4907, 99, 103
NGC 4908, 99
NGC 4911, 99, 103
NGC 4919, 99, 103
NGC 4921, 99, 103
NGC 4923, 99, 103
NGC 4926, 99
NGC 4926A, 99
NGC 4927, 99
NGC 4956, 71
NGC 5002, 75
NGC 5005, 68, 72, 73, 75
NGC 5014, 72, 75
NGC 5023, 72
NGC 5033, 68, 72, 75
NGC 5055, 72
NGC 5074, 72
NGC 5103, 72
NGC 5112, 72
NGC 5123, 72
NGC 5141, 72
NGC 5142, 72
NGC 5149, 72
NGC 5169, 62, 64, 66
NGC 5173, 62, 64, 66, 72
NGC 5194, 72
NGC 5195, 29, 62, 72
NGC 5198, 62, 66, 72
NGC 5229, 64, 65, 66
NGC 5256, 66
NGC 5272, 72
NGC 5273, 72
NGC 5290, 72
NGC 5297, 72
NGC 5301, 72
NGC 5313, 72
NGC 5318, 72
NGC 5320, 72
NGC 5326, 72
NGC 5336, 72
NGC 5337, 72
NGC 5347, 72
NGC 5350, 28, 72, 75
NGC 5351, 72
NGC 5353, 28, 75
NGC 5354, 28, 75
NGC 5355, 75
NGC 5358, 75
NGC 5363, 25, 29
NGC 5364, 25, 29

NGC 5371, 75
NGC 5375, 72
NGC 5377, 72
NGC 5380, 72
NGC 5383, 72
NGC 5394, 26, 29, 72
NGC 5395, 26, 29, 72
NGC 5406, 72
NGC 5426, 25, 29
NGC 5427, 25, 29
NGC 5440, 72
NGC 5544, 72
NGC 6166, 84
NGC 6194, 106
NGC 6196, 106, 107
NGC 6197, 106, 107, 108
NGC 6199, 106, 107
NGC 6207, 106, 107, 108
NGC 6231, 55
NGC 6822, 6, 9, 10, 11, 82
NGC 6946, 15, 17, 18, 19
NGC 6951, 15, 19
NGC 7217, 15, 19
NGC 7314, 19
NGC 7320, 79
NGC 7325, 78, 79, 80, 81
NGC 7326, 78, 79, 80, 81
NGC 7327, 78, 79, 80, 81
NGC 7331, 15, 19, 76, 78, 79
NGC 7333, 78, 79, 80, 81
NGC 7335, 76, 78, 79, 80, 81
NGC 7336, 76, 78, 79, 80, 81
NGC 7337, 76, 78, 79, 80, 81
NGC 7338, 79, 81
NGC 7340, 79, 81
NGC 7410, 15, 19
NGC 7448, 19
NGC 7457, 17, 19
NGC 7469, 19
NGC 7479, 14, 17, 19
NGC 7619, 19, 21
NGC 7625, 19
NGC 7640, 17, 19
NGC 7679, 19
NGC 7682, 19
NGC 7714, 5, 21, 15, 19
NGC 7723, 19, 21
NGC 7742, 19
NGC 7793, 15, 19
NGC 7814, 15, 19
93 Leonis, 89
Noonan, Thomas, 98
Normal spirals, 15
Novae, 54, 90
O III filter, 9, 37
Oort, Jan H., 105
Open clusters, 70, 71, 92
Orion, 96
Orion Nebula, 36
Palomar Mountain Observatory, 58, 98
Palomar Obervatory Sky Survey, 30, 87, 106
Paris Observatory, 62
Paternostro, Mark, 55
Paturel, Georges, 97
Peach, J. V., 105
Peculiar galaxies, 15, 21,1 9, 26, 29
Pegasus, 10, 14, 15, 17, 19, 78
Pegasus dwarf galaxy, 9, 12
Pellet, A., 61
Perseus, 19, 55
Perseus galaxy cluster, 84, 98
Pi[5] Orionis, 96
Pisces, 10, 19, 21
Planetary nebulae, 59
Polakis, Tom, 6, 11, 12, 30, 34
Population II stars, 93
Potter, Ron, 59, 62
Princeton University, 84
Princeton University Observatory, 60
Priser, John B., 101, 103
Publications of the Astronomical Society of the Pacific, 12
Radloff, Max, 68
Reaves, Gibson, 10
Red supergiants, 54
Regulus, 13, 82
Regulus System, 10
Revised New General Catalogue, 64, 106
Ring galaxies, 21
Roberts, Isaac, 90
Roberts, Morton, 54, 61
Rood, Herbert J., 84, 89
Rood-Sastry Scheme, 85

Rosino, L. 54
Rosse, Lord (William Parsons), 15, 79, 80, 81
Rubin, Vera C., 54, 61
S Andromeda, 90
Sag DIG, 10, 12
Sagittarius, 6, 9, 82
Sandage, Allan, 6, 36
SAO 65569, 109
Sastry, K. L. V. N., 84, 89
Schade, D., 61
Schultz, Herman, 79, 80
Schur, Chris, 13
Schwarzchild, M., 60, 61
Scorpius, 55
Sculptor, 10, 15, 16, 17
Sculptor group, 68
Sculptor System, 10
Second Reference Catalogue of Brght Galaxies, 105
Sextans, 10, 25
Sextans A, 6
Sextans Dwarf, 13
Seyfert galaxy, 21, 26
Shane, C. Donald, 98, 105
Shapley, Harlow, 98, 91
Sharov, A. S., 54, 61
Simien, F., 61
Skalnate Pleso Atlas Catalogue, 19
Skiff, Brian, 6, 33, 36, 37, 54, 56, 93, 98, 100, 102, 105
Sky Atlas 2000.0, 6
Slipher, Vesto M., 90, 91
Small Magellanic Clouds, 10, 22
Sombrero Galaxy, 15
Spiral arms, 15, 55, 75
Spiral galaxies, 55, 99, 54
"Spiral nebulae", 78, 86, 91, 98
Star clouds, 55
Star clusters, 6
Stellar associations, 55, 92
Stephan's Quintet, 15, 79
Sulentic, J., 79, 81
Supernovae, 90
Surface brightness, 12, 15, 17, 21, 30, 54, 70, 82, 86, 103
Swift, Lewis, 65
Tammann, Gustav, 6
Tarantula Nebula, 37
Tempel, W., 79, 80, 81
10 Orionis, 96
Third Reference Catalogue of Bright Galaxies, 97
Thompson, Laird A., 98, 105
Tifft, W., 79, 81
3C 273, 27
Triangulum, 17, 19, 21, 25, 54, 82
Tucana, 10
Tucana Dwarf, 13
Tully, R. Brent, 30
U. S. Naval Observatory, 101, 103, 99
UCLA, 98
UGC 4305, 36
UGC 4459, 35, 36, 37
UGC 5139, 36
UGC 5423, 36
UGC 5666, 36
UGC 6456, 36, 37
UGC 6680, 104
UGC 6697, 104
UGC 6719, 104
UGC 6725, 104
UGC 7175, 70, 71
UGC 7699, 71
UGC 8201, 36, 37
UGC 8303, 75
UGC 08320, 66
UGC 08331, 66
UGC 08468, 66
UGC 08470, 66
UGC 08485, 66
UGC 08499, 66
UGC 08538, 66
UGC 8538, 65, 67
UGC 08550, 66
UGC 08588, 66
UGC 8588, 65, 67
UGC 08597, 66
UGC 08601, 66
UGC 08611, 66
UGC 08632, 66
UGC 10447, 109
UGC 10477, 108
UGC 10544, 108, 109
UGC 10566, 108, 109
UGC 10567, 107, 108, 109
UGC-A86, 10

UMa I cloud, 68
University of Brunswick, 84
University of Texas, 80, 107
University of Washington, 54
Upgren 1, 70, 71, 75
Uppsala General Catalogue of Galaxies, 64, 105, 106
Uranometria 2000.0, 12, 62, 64, 65, 67, 68, 83, 87, 88, 96, 106, 107, 108, 109
Ursa Major galaxy cluster, 75
Ursa Minor, 10
Ursa Minor dwarf, 12, 13
Van den Bergh, Sidney, 61, 55
Van Maanan, Adrian, 91, 92
Vaughn, Chuck, 33
Vetesnik, M., 61
Viale, A., 61
Virgo, 25, 29, 58, 68, 84
Virgo Cluster of galaxies, 6, 10, 22, 27, 29, 84, 98
Vorontsov-Velaminov, B.A., 65
VV 21, 29
VV 30, 27
VV 48, 29
VV 188, 26
VV 206, 29
VV 209, 25
VV 246, 26
VV 189, 25
Walker, Rich, 12
Warner Observatory, 65
Warren Rupp Observatory, 62, 106, 107
Whirlpool Galaxy, 22
Whitehurst, Robert, 54, 61
Williams, B. A., 87
Wirtanen, Carl, 98
WLM, 9, 10, 11
Wolf, Max, 9, 98
X-ray sources, 99
Yahil, Amos, 6
Z421-15, 97
Z421-16, 97
Z421-17, 97
Z421-18, 97
Z421-18a, 97
Z421-18b, 97
Z421-19, 97
Zeta[1] Scorpii, 54
Zussman, Kim, 14, 18
Zwicky 97.101, 104
Zwicky 97.102a, 104
Zwicky 97.102b, 104
Zwicky 97.110, 104
Zwicky 97.112, 104
Zwicky 97.124, 104
Zwicky 97.125, 104
Zwicky 97.131, 104
Zwicky 97.74, 104
Zwicky 97.83, 104
Zwicky 97.85, 104
Zwicky 97.86, 104
Zwicky 97.90a, 104
Zwicky 97.90b, 104
Zwicky 97.92, 104
Zwicky 97.99, 104
Zwicky 127.40, 104
Zwicky 127.47, 104
Zwicky 245.036, 66
Zwicky 245.038, 66
Zwicky 246.002, 66
Zwicky 246.003, 66
Zwicky 246.004, 66
Zwicky 246.005, 66
Zwicky 246.006, 66
Zwicky 246.007, 66
Zwicky 246.010, 66
Zwicky 246.011, 66
Zwicky 246.013, 66
Zwicky 246.016, 66
Zwicky 246.018, 66
Zwicky 246.020, 66
Zwicky 246.021, 66
Zwichy 197.012, 108, 109
Zwicky 127-109, 87
Zwicky 127-133, 87
Zwicky 197.013, 108, 109
Zwicky 197.016, 107, 108, 109
Zwicky, Fritz, 66, 80, 98